AMMUNITION

AMMUNITION

SMALL ARMS, GRENADES AND PROJECTED MUNITIONS

Ian V. Hogg

This book club edition is manufactured under license
from Greenhill Books/Lionel Leventhal Limited, London.

ISBN 1-85367-323-4

Manufactured in USA

Contents

Preface

It must always be remembered that ammunition is the weapon. There are numerous ways of delivering that weapon – rifles, machine guns, mortars, aircraft, warships, even throwing, but they are simply the final stage in transportation from the munitions factory to the enemy. Without ammunition all these 'weapons' are useless.

Ammunition design tends to respond to tactical demands. The need to hide the movement of troops created smoke shells, tanks created anti-tank ammunition and so on. As a result, ammunition design and development goes in cycles of particular activity, according to what happens to be seen as the predominant threat at the time. At the present time most of the activity seems to be in the field of long-range delivery of sub-munitions by artillery, but within the next decade this will undoubtedly have changed and something else will be making demands on the designers' time.

In spite of its importance, ammunition is nevertheless a neglected field when it comes to the provision of information. There are innumerable books about 'weapons' but few about 'the weapon', with the exception of small arms ammunition. This subject is covered largely due to the hand-loading and collecting aspects, activities which have little correspondence in other ammunition areas.

This book, therefore, is not intended to be a complete catalogue of ammunition; such a catalogue would demand several hundreds of pages and involve a great deal of repetition. It is intended to indicate the scope of ammunition, its versatility and the current trends in design, and how they have evolved from earlier models. It also recites a few of the basic factors underlying ammunition in order to make things understandable for people who have not, hitherto, given much thought to this side of the study of weapons. It also illustrates the lengths to which human ingenuity can go in the struggle to win wars.

A word of warning: nothing in this book should be taken to encourage anyone to attempt to dismantle ammunition. Ammunition is put together in a manner such as to discourage idle curiosity and violent attempts to uncover the mysteries usually end in disaster. Ammunition is blind. It does not recognise uniforms or faces; it simply operates when sufficiently stimulated and it cares not who does the stimulating.

The author and publisher wish to thank the many manufacturers and agencies who have kindly provided illustrations and drawings for inclusion in this book. The information and data given herein is that which has been publicly released by the various manufacturers, and where items of information have not been so released they have been omitted from the various tables of data.

Small Arms and Cannon Ammunition: Small Arms

'Small Arms' are generally defined as weapons of a calibre up to about 15mm (0.60in) and capable of being operated by one or two men; 'Cannon' means the group of automatic weapons of 20mm to 30mm calibre. The ammunition of these two groups is similar in its general appearance and can be conveniently considered together.

All ammunition is designed with an eye to its tactical effect; you must know what you want to do before you can design something to do it. The tactical purpose of small arms ammunition is, generally, its anti-personnel effect against the enemy's troops. This means that the majority of small arms ammunition – that for pistol, rifles, sub-machine guns and machine guns – is the plain 'ball' round, a solid bullet designed purely to kill or wound. This sounds simple enough, but it is complicated by the restrictions placed on bullet design by the various international agreements, from the St Petersburg Declaration to the Hague and Geneva Conventions, which rule that 'explosive' bullets (by which is meant bullets which easily deform or fragment on impact and thus cause excessively severe wounds) are forbidden. And, strangely enough, this is more or less universally adhered to; soft-point and hollow-point bullets, or bullets of pure lead, are never seen in military equipment. I say 'more or less' because although these types are proscribed, there is a constant, if low priority, search for a bullet which obeys the rules and yet delivers severe wounds and instant lethality. The thing which keeps people on the straight and narrow path of rectitude is the thought that if they stray, so will the enemy, and he might have something worse than you have. Small arms ammunition was the first of the mutually assured deterrents.

The 'solid' ball bullet is not, in fact, solid, for a variety of reasons. In the first place, bear in mind that the instant the bullet leaves the barrel of the weapon, it is 'free-wheeling'; the impulse which drives it has stopped and it is relying upon its own momentum. For this reason it needs to be heavy; a light bullet soon loses its velocity and striking power. Lead would be ideal, but, as noted above, solid lead is forbidden because it deforms easily on impact. Lead also leaves deposits in the bore as the bullet passes through, which soon render the weapon inaccurate and if not quickly removed, will prevent it firing at all. Steel would not be heavy enough and moreover steel would wear the inside of the weapon's barrel away due to friction. All this was examined closely in the 1880s; the French solved it by adopting a bullet of solid copper. But a Swiss officer, Major Rubin, developed the 'compound bullet' which has, with variations, been the normal type ever since.

The compound bullet consists of three elements: the core, the jacket and the envelope. The jacket is a hollow steel shape, pointed at the front and open at the rear. The core is a slug of lead which fits inside the jacket, the rear end of which is then turned in to keep the core in place. Since the steel jacket would wear away the gun barrel, it is now enclosed in a soft copper or gilding metal envelope which bites into the rifling, but does not generate the same wear as would naked steel.

The core may be pure lead, or it may be part lead and part steel, or part lead and part plastic, or part steel and part aluminium – the combinations are endless and the object in view is the get the centre of balance of the bullet in the right place to ensure stable flight whilst still having the desirable heavy core to give the bullet its momentum. Moreover there is today a desire to defeat body armour and steel helmets and thus most rifle and machine gun bullets now have a core, or part core, of steel.

The designer's task is also bedevilled by the constant demand for smaller rifles and smaller calibres. Until about 1875 the normal calibre for the infantryman's rifle was in the order of half an inch or so, but it then began to reduce. This was the start of the theory that increased velocity would compensate for loss of weight in the bullet, since the momentum and thus the kinetic energy in the bullet would remain the same or even increase. There was also, no doubt, the question of the economics of the price of lead, after somebody had calculated that during the American Civil War the expenditure of ammunition amounted to something like three tons of lead being shot off for every man killed. The calibres of rifles began to reduce, until in the late 1880s the French produced an 8mm rifle and an effective bullet, propelled by smokeless powder at a (for the time) high velocity. By the turn of the century calibres of 6.5mm to 8mm were the normal throughout most armies, together with compound bullets and the situation remained like this throughout World War Two.

During the course of World War Two, however, various people began having second thoughts about the infantryman's rifle. The design had stemmed from the tactical demands which had been apparent in the late 19th century and the view at that time had been conditioned by the Franco-Prussian war in which, for the first time, the soldier's rifle had much the same range as did his supporting artillery, with the result that a hail of long range rifle fire frequently devastated enemy artillery while it was attempting to get into position. Long range fire was thus very desirable and this demanded a heavy bullet, a powerful cartridge and a long and heavy rifle. The South African War backed up this demand, because the British troops were attacked by long range fire from Mauser rifles wielded by Boer farmers. The British rifles were capable of ranging just as far but not, apparently, with the same degree of accuracy. This led to demands for a better and more powerful rifle, but what the critics failed to appreciate was that the Boer farmer, a frugal man, had been taught since boyhood to hunt for the pot and do it with the minimum of ammunition. A man who could bring down a running antelope at 300 yards was going to have very little difficulty in hitting a standing man at 500 yards or more. Ammunition was not the key factor, rather it was training and skill.

By the late 1930s questions were being asked as to whether this long range capability was really necessary in the modern battle. How many times did a soldier get a clear shot at an enemy more than 400 yards away? Indeed, how many soldiers could even see a camouflaged enemy, in muddy drab clothing and crouching under a hedge at 400 yards or more? So why have a cartridge capable of shooting to 2,000 yards or more when such performance was never needed? Moreover, there was keen competition among armies to bring a reliable automatic rifle into service, one with the capability of firing like a light machine gun in an emergency. Rifles firing the old style full power cartridge were uncontrollable at automatic fire, the recoil forces being too much for the soldier to handle.

Thus the 'short' cartridge and the assault rifle came into existence; a bullet of the normal calibre, but shorter and lighter, with a cartridge case considerably shorter and a lighter charge of powder. A man could carry more ammunition, the rifle could be shorter and lighter, it would sustain automatic fire and it performed perfectly satisfactorily up to about 600 yards range, further than the soldier was ever likely to need to fire.

In the late 1950s, possibly as a result of some rather dubious battlefield interviews carried out in the latter part of World War Two and in Korea, the US Army decided that its soldiers were not shooting well enough. They therefore instituted a programme to determine methods of improving the weapon so as to overcome the shortcomings due to lack of training and skill in shooting. 'Project Salvo' investigated several ideas, including shooting two bullets out of one cartridge, shooting needle -like darts singly or in clusters, shooting a short burst of fire for every trigger pull and so on. The US Army also investigated the use of even smaller calibres and cartridges to produce lighter rifles with minimal recoil so that the soldier would not flinch as he pulled the trigger and the weapon would not hit him severely in the shoulder no matter how carelessly he held it. Calibres went down and down, until work was being done on .17 calibre bullets – which is fractionally smaller than that of the average air pistol. The expert's view was that at high velocity this small calibre would deliver the same blow as a heavier bullet.

Eventually the world settled on 5.56mm (.223 inch) as the standard calibre (apart for the Russians who selected 5.45mm, which bears favourable comparison). With the 5.56mm bullet came fearsome tales of its deadly potential – strike a man in the arm and the pressure wave would paralyse his heart etc – much of which was accepted for several years until the investigation of wound ballistics became a scientific discipline rather than a by-product of military surgery.

Now the pendulum is starting to swing back. The wound ballistics experts have pointed out that the modern small calibre bullet is not as effective at disabling a soldier as were the heavier bullets of yesterday. Soldiers who have had combat experience with both types of bullet agree that whereas a charging man will be stopped with one or two shots with an 8mm bullet, four or five with 5.56mm bullets will not necessarily stop him. Should the range get much beyond 400 yards the light bullet runs out of steam and what is more serious, tends to lose its accuracy. So the Russians are now looking into a 6mm bullet; not a large improvement perhaps but a significant move towards increasing the calibre.

There have, in the past, been many Boards, Commissions and Panels convened to research into the question of the ideal calibre and to submit reports. Almost without exception, every one of the boards has reported that a 7mm bullet is the ideal. Only once however has anyone taken any action on it: the British Army adopted a 7mm (.280) cartridge in 1949, only to reject it a few months later for political reasons – NATO would not accept it as their standard. Should the Russians make their point with the 6mm bullet, there is a possibility, after a 60 year interval, of a return to the 7mm calibre cartridge in time for the next generation of rifles.

While designing a ball bullet is difficult, there are additional problems presented by designing those two essential auxiliaries, armour-piercing and tracer bullets, in the current small calibre. The core of the armour-piercing bullet is of specially hardened steel or, in some designs, a tungsten compound. It must be sheathed in lead in order to allow the jacket to deform as it enters the rifling, but the weight must be the same as, or very close to, that of the ball bullet, so that the trajectory is the same and the sights of the weapon are effective for both types of bullet. When one considers that the ball bullet is expected to defeat a standard steel helmet at 600 metres range, one wonders how much improvement can be expected of an armour-piercing bullet in such a small size.

Tracer is another problem area. Into a small 5.56mm bullet has to be placed a quantity of pyrotechnic material which will burn with sufficient brightness and colour to make the trajectory visible. Furthermore, since it is undesirable to have the bullet leave the barrel with the tracer burning, as this tends to both dazzle the firer at night and reveal the position of the weapon to enemy observers, there must be a thin layer of delay mixture which will burn through before igniting the tracer when the bullet is about 100 yards from the muzzle. The tracer will then burn out at about 500 yards range. Considering that the entire flight of the bullet takes less than a second, it is apparent that the amount of delay compound and tracer mixture becomes very critical so as to ensure that everything takes place at the right time. An additional point which is frequently overlooked is that as the tracer composition burns away, so the weight and balance of the bullet change and thus the tracer bullet will not follow exactly the same trajectory as the ball and armour-piercing bullets. So the tracer is designed in such a way that the point of impact will be the same for all types of bullet at some specific range, normally chosen as being the most likely combat range of the particular weapon.

Generally, ball, armour-piercing and

tracer are the types of bullet in most common use. There are occasions, though, when other types are necessary and this is particularly the case with the heavier types of machine gun. These can be used with good effect against vehicles and other material targets and it makes sense to develop types of bullet which can enhance the terminal effect against these types of target. One example is the armour-piercing-incendiary bullet widely used with .50 and 12.7mm machine guns. The object here is to defeat lightly armoured vehicles, such as armoured cars and light personnel carriers and, by penetrating the fuel tanks or engine compartment with the armour-piercing-incendiary bullet, set the vehicle on fire. Ordinary armour-piercing with tracer could perhaps have the same effect, but tracer cannot be relied upon to provide ignition. The usual design holds a small quantity of an incendiary composition in the tip, beneath the jacket and in front of the armour-piercing steel core. The bullet strikes, the jacket crumples and the intense heat generated by the impact ignites the incendiary mixture while the core continues its path through the target, sucking the burning incendiary mixture in behind it.

In spite of the restrictions of the various protocols, explosive bullets are used in special circumstances. The particular case which most commonly demands such a bullet is the difficulty in observing strikes at long range when using armour-piercing or ball bullets. An explosive bullet will provide a distinct flash and explosion on impact, thus positively announcing a hit and confirming that the inert bullets being fired at the same range are doing their work. For this reason they are frequently called 'observation' or 'ranging' bullets. Their most common use today is with aiming rifles attached to anti-tank weapons; the bullet is carefully designed to duplicate the trajectory of the parent weapon and single shots are fired, with varying points of aim, until a strike is seen, whereupon the parent weapon is fired with the same aiming information.

Although the various bullet types just described perform almost everything necessary on the battlefield, there is a constant research effort being expended to improve them wherever possible. The difficulty lies in the physical dimensions of the ammunition; there is a limit to what can be achieved within the constraints of any particular calibre. There are two tendencies worth noting however, both in the heavy machine gun field. The first has been the development of extremely powerful explosive armour-piercing bullets, virtually small shells, for the .50 Browning machine gun. These were developed by the Raufoss Arsenal in Norway some years ago and the elements of the design have been widely adopted by other countries. This 'NM140' bullet incorporates penetrative, incendiary and fragmentation effects; it will defeat 16mm of armour at 400 metres range and after doing so it will explode and produce about 20 effective fragments inside the target. It also produces a shower of incendiary particles which are still effective up to 15 metres behind the target plate.

The second is the revival of the double bullet round. This idea has cropped up more than once during the 20th century, most particularly during the American 'Salvo' project when rounds in 5.56mm and 7.62mm calibre were produced having one bullet in the mouth of the case in the usual way and a second bullet inside the case. This second bullet had the flat base cut slightly at an angle, so that as it left the muzzle the gases behind it would escape at one side of the bullet first and thus cause it to diverge slightly from the trajectory of the first bullet. In this way it was hoped that should the first bullet miss its target by reason of an aiming error, then the second might well score a hit, due to its divergence. The idea was tried in Vietnam, with questionable effect and was then forgotten. The concept has been revived in Russia in a new cartridge for 12.7mm heavy machine guns and it remains to be seen whether it achieves any success.

The following material deals with military cartridges in current use by various countries and also one or two commercially developed cartridges which are being, or may be used, in a military application.

4.7mm OH DE11

OH stands for 'ohne Hülse' or 'without case' and this is the caseless cartridge developed by Heckler & Koch for their revolutionary G11 rifle; DE11 is the German army nomenclature for the round. Heckler & Koch were assisted by Dynamit Nobel in the development of this cartridge from about 1970 onwards. The first design used compressed smokeless powder, with the bullet embedded in one end and a combustible cap in the other. This worked satisfactorily, but tended to 'cook off' if loaded into a hot chamber and not fired immediately. More work was carried out and an entirely new propellant based upon a denatured high explosive was developed which lifted the 'cook off' temperature by about 100°C and removed the hazard.

The final design has the bullet set entirely into a shaped block of propellant, again with a combustible cap at the rear end. In front of this is a small initial charge of powder which, when the cap fires, blows the bullet out of the block of propellant and into the rifling. During this phase the propellant is shattered by the explosion of the initiatory powder and ignited. It then explodes and drives the bullet from the barrel. This sequence is necessary in order to have the bullet firmly seated and engraved into the rifling before full pressure is applied, thus preventing any escape of gas around the bullet which would lead to erosion of the bore and erratic performance. The bullet itself is of conventional pattern, a gilding metal envelope surrounding a steel jacket and lead core. It does not usually fragment upon impact and will penetrate a standard steel helmet at 600 metres range.

DATA

Overall length 1.29in (32.80mm)
Case length n/a
Rim diameter 0.311in (7.90mm) square
Bullet weight 49.3grains (3.20g)
Muzzle velocity 3051ft/sec (930m/sec)
Muzzle energy 1,021ft/lbs (1380J)

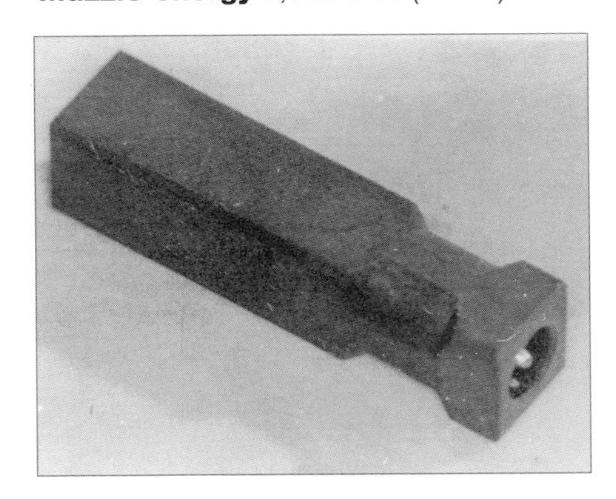

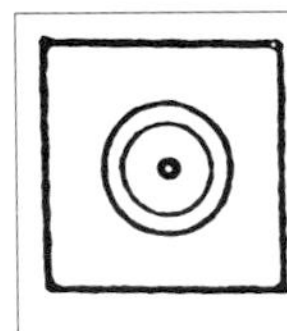

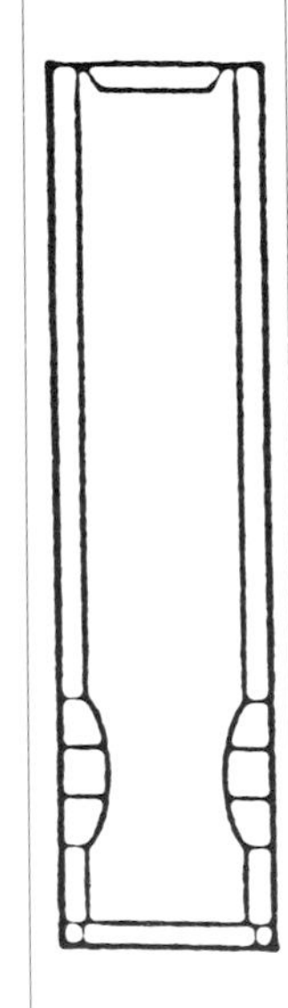

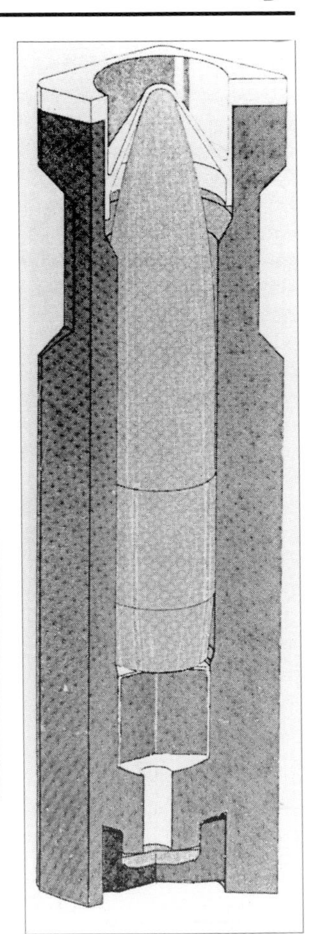

5.45 x 18mm Pistol

Russia

What lay behind the development of this cartridge is not known, nor may ever be, apart from the Russian requirement to have a cartridge which is incapable of being fired in anyone else's weapons. It was developed during the 1970s, for the PSM pistol, a small blowback weapon issued to security police and troops. A bottle necked round, it flies in the face of received wisdom, which states that a necked case should never be used in a blowback weapon since it tends to lead to sticking cases and erratic ejection. The pistol appears to work, but why it was thought necessary to go to such lengths to fire a 40 grain bullet at 1,030ft/sec remains a mystery. The bullet is ogival, flat-tipped, full jacketed and is abnormally long at 14.5mm or 2.8 calibres. The gilding metal jacket covers a two part core, the front end being of steel and the rear end of lead; this is an unusual construction for a pistol bullet, though it is very similar to the 5.45mm rifle bullet. It is claimed to have enormous penetrative powers against body armour. The cartridge is only manufactured in Russia and only ball bullets are known.

DATA
Overall length 0.980in (24.90mm)
Case length 0.700in (17.80mm)
Rim diameter 0.297in (7.55mm)
Bullet weight 40.1 grains (2.60g)
Muzzle velocity 1,033ft/sec (315m/sec)
Muzzle energy 95ft/lb (129J)

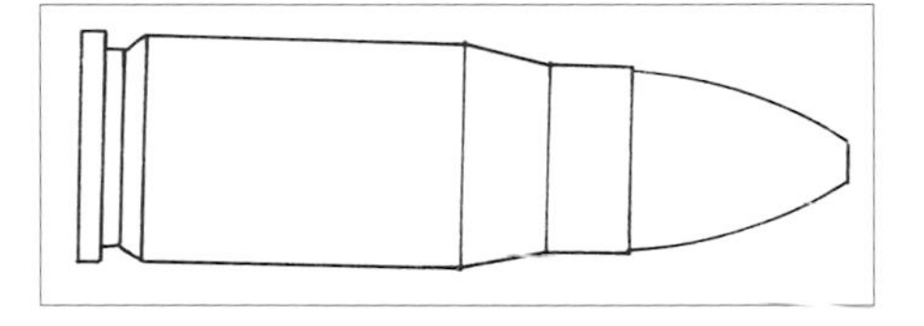

(Below) The 5.45mm Soviet pistol round compared with 9mm Short, 9mm Parabellum and .45ACP rounds.

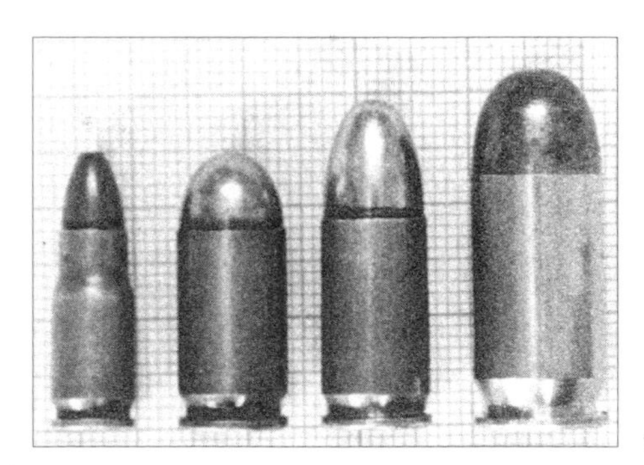

5.45 x 39.5mm Rifle Russia

Russia's answer to the 5.56mm M193 cartridge, adopted in 1974 though specimens were not seen in the west until 1979. A rimless, bottle necked round, it generally resembles the 5.56mm cartridge but is slightly shorter and somewhat fatter. Cases are of lacquered steel, with a Berdan primer and two fire holes. The bullet is of compound construction; the gilding metal jacket covers a rear core of mild steel, with a short front core of lead and a hollow space in the bullet tip. This tends to collapse on impact and cause the bullet to take up a curved path, producing a severe wound cavity instead of a simple penetration.

The cartridge is now manufactured in Germany, Romania, Poland, Serbia and Bulgaria as well as in Russia. Standard bullets are the 53 grain ball, giving a muzzle velocity of 2,950ft/sec, red tracer, armour-piercing and a plastic practice ball.

DATA
Overall length 2.224in (56.50mm)
Case length 1.555in (39.50)
Rim diameter 0.393in (10.00mm)
Bullet weight 53 grains (3.435g)
Muzzle velocity 2952ft/sec (900m/sec)
Muzzle energy 1,023ft/lb (1,383J)
Velocity at 300 metres 2,060ft/sec (628m/sec)
Energy at 300 metres 500ft/lb (677J)

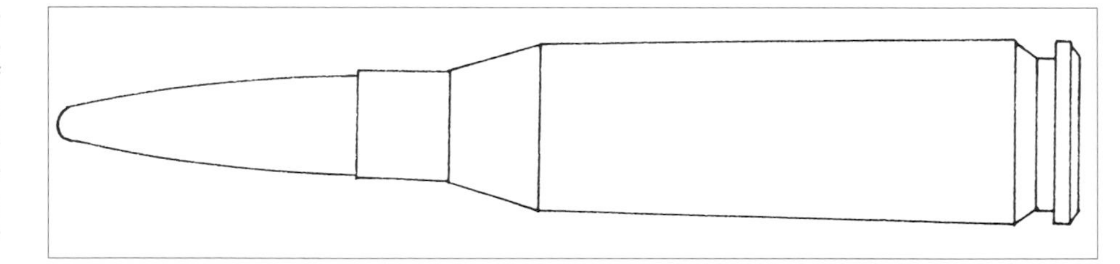

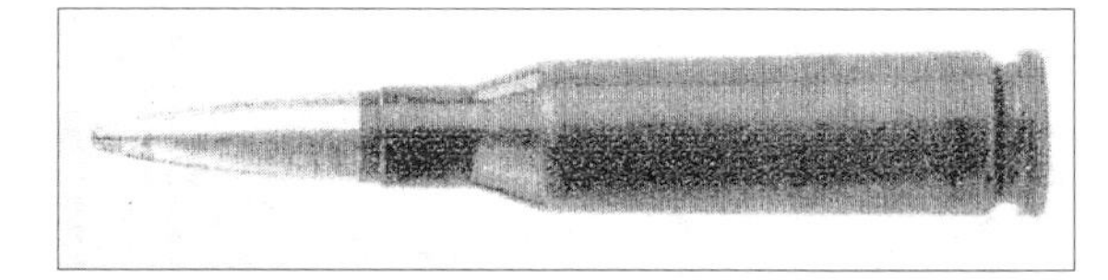

5.56 x 45mm NATO

This is the cartridge which emerged from the 1978–81 NATO Small Arms Trial as the winner for the position of standard NATO infantry rifle round. Although there were a number of exotic contestants, it was fairly obvious from the start that the 5.56mm cartridge would be selected: the Americans had unilaterally adopted it as their infantry round, several nations had begun manufacturing rifles in this calibre and there was a considerable amount of money invested in ammunition production. But instead of simply rubber stamping the American round, the trials went into great detail and eventually selected the SS109 round developed by Fabrique Nationale Herstal. This uses a somewhat heavier bullet than the original American round and launches it at a slightly lower velocity, but its down range performance is better owing to its better carrying power.

Current types available are ball, tracer and armour-piercing, though in this calibre one should not expect too much from the armour-piercing bullet.

DATA
Overall length 2.256in (57.3mm)
Case length 1.75in (44.45mm)
Rim diameter 0.374in (9.50mm
Bullet weight 62 grains (4.00g)
Muzzle velocity 3,051ft/sec (930m/sec)
Muzzle energy 1,264ft/lb (1,708J)
Velocity at 300 metres
2,133ft/sec (650m/sec)
Energy at 300 metres 617ft/lb (834J)

(Below) A US 5.56mm Duplex round, cut open to show the second bullet.

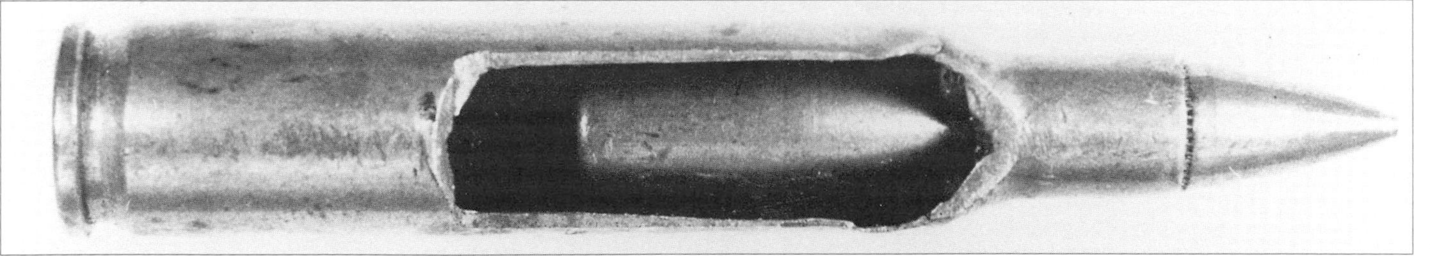

5.7 x 28mm P90 — Belgium

This was developed in the mid-1980s by Fabrique Nationale of Herstal, Belgium, for their innovative P90 Personal Defence Weapon. Their object was to produce a weapon with which to arm that two-thirds of an army which does not require an assault rifle for their day to day business and, in Fabrique Nationale's own words, 'To introduce a new calibre which will advantageously replace the obsolescent 9 x 19mm Parabellum NATO ammunition'.

The round resembles a rifle round more than a submachine gun round and its performance is far in advance of any current submachine gun, the ball bullet being capable of piercing 48 layers of Kevlar at over 50 metres range or a steel helmet at 150 metres. The bullet has a synthetic core and steel jacket and does not fragment on impact, though it has excellent stopping power. A hard core discarding sabot bullet has also been developed which is capable of defeating standard NATO armour plate targets.

At present the ammunition is made only by Fabrique Nationale Herstal, but with the increasing acceptance of the P90 weapon and with the likelihood of the round being NATO standardised, it can be expected to be licensed for manufacture elsewhere in the future.

DATA
Overall length 1.711in (43.45mm)
Case length 1.137in (28.90mm)
Rim diameter 0.307in (7.80mm)
Bullet weight 23 grains (1.50g)
Muzzle velocity 2,788 ft/sec (850m/sec)
Muzzle energy 400ft/lb (540J)

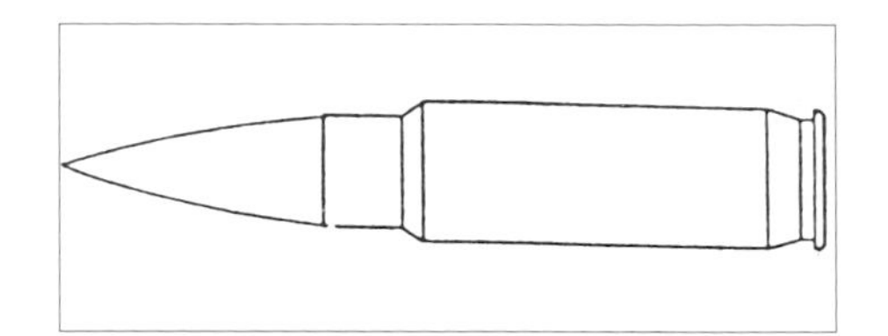

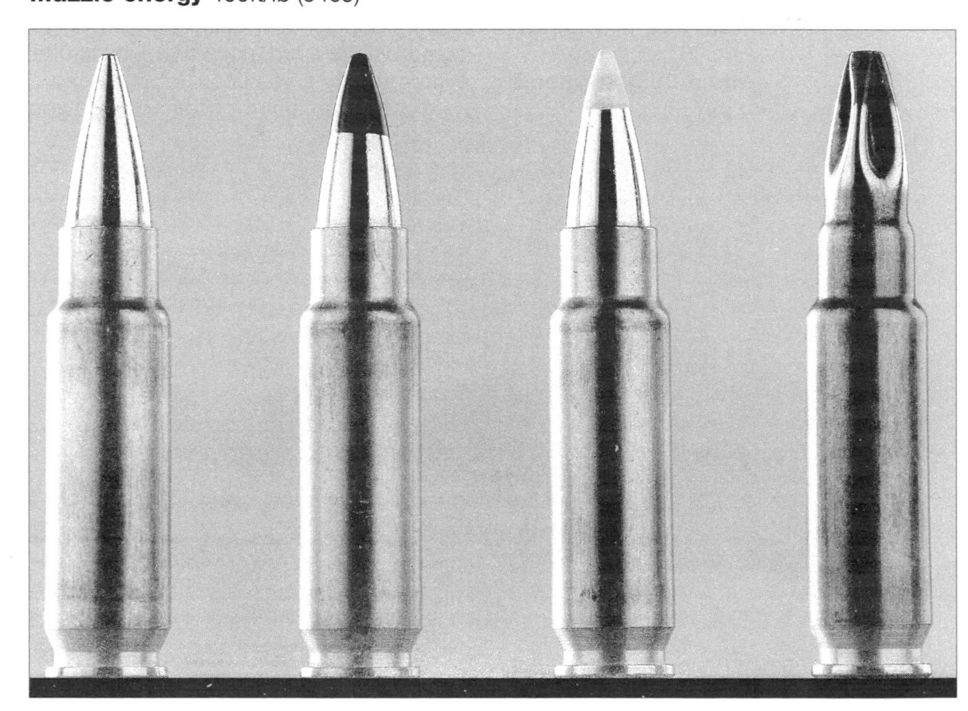

7.62 x 39mm

Russia

This is the well known Soviet M1943 short-case round. The Soviets had been experimenting with short cartridges prior to WW2 and their discovery of the German 7.92 Kurz cartridge and its associated assault rifle spurred them to return to this project in 1943 and develop this cartridge. It was handed to Simonov, one of their most prominent weapon designers, and he modified an existing design of automatic rifle to suit the new round. This was tested in action in 1944, some minor defects were noted and corrected and in 1945 the Simonov SKS carbine was adopted as the future standard Soviet army rifle. Also in 1944 Kalashnikov developed his first automatic rifle, a wood-stocked carbine using gas operation and a rotating bolt, but in view of the adoption of the Simonov this design was turned down. Kalashnikov then went on to develop the well known AK series of rifles, with the result that the M1943 cartridge has become known throughout the world.

The standard Soviet ball bullet is a 123 grain conventional streamlined type with steel core, steel jacket and gilding metal envelope; ball bullets produced in other countries may be non-streamlined and have lead cores, but are adjusted to the same weight as the Soviet standard. Other available types include the usual tracer, armour-piercing, armour-piercing-incendiary and incendiary ranging bullets. Cases are usually of lacquered steel, though brass and brass coated steel are commonly made by producers outside Russia. The proliferation of Russian weapons around the world has led to this cartridge being made in western countries as well as in former Warsaw Pact countries.

DATA
Overall length 2.197in (55.80mm)
Case length 1.522in (38.65mm)
Rim diameter 0.445 (11.30mm)
Bullet weight 123 grains (7.97g)
Muzzle velocity
2,300ft/sec (710m/sec)
Muzzle energy 1,487ft/lbs (2,010J)
Velocity at 300 metres
1,545ft/sec (471 m/sec)
Energy at 300 metres
1,195ft/lb (884J)

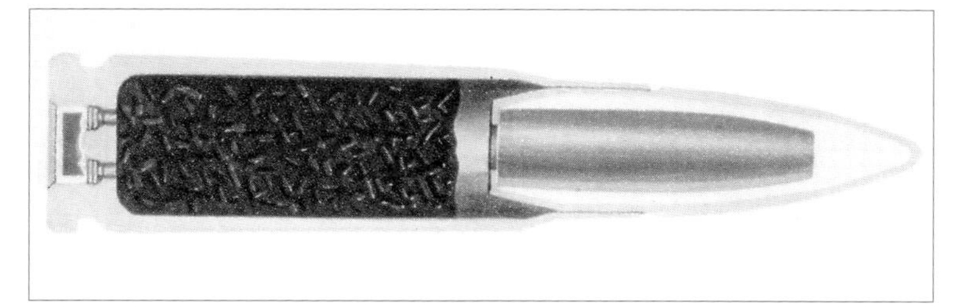

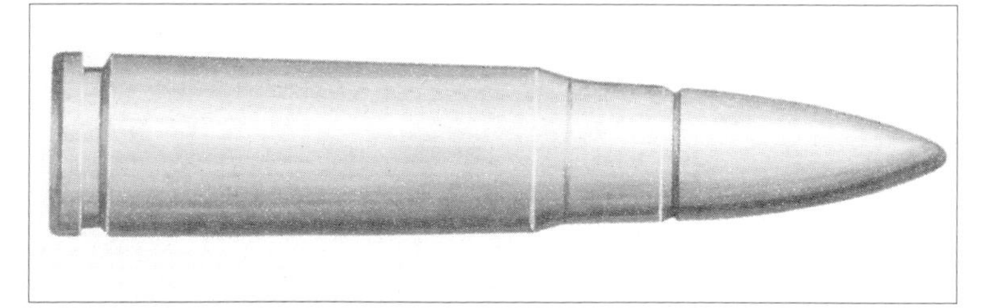

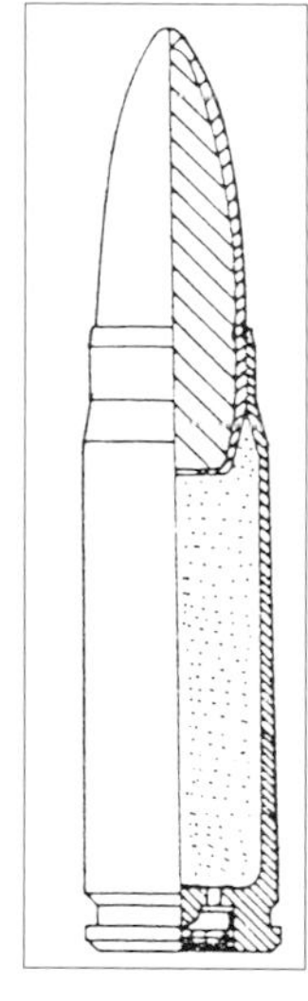

7.62 x 51mm NATO

NATO's standard from the 1950s to the 1980s, this initially was the American T65, a compromise put forward in opposition to the British 7mm (.280) round in the late 1940s. At that time the idea of a short cartridge and lighter rifle held no appeal for the US Army, whilst the rest of NATO balked at the full power .30-06 Springfield cartridge. The compromise was reached by shortening the .30-06 case and putting a lighter bullet into it, creating a round which was neither full power nor 'compact' and putting paid to any European ideas of adopting short assault rifles for several years. A great deal of hard work in the ammunition factories was required before the cartridge managed to gain a reputation for accuracy.

The ball bullet is NATO standardised as a streamlined pattern weighing 149 grains and with a lead/antimony core, steel jacket and gilding metal or copper envelope. The case is rimless, bottle necked, Berdan primed and made from brass or lacquered steel. The usual armour-piercing and tracer are also standardised, but various manufacturers produce a wide variety of ball designs with different bullet weights, specially selected sniper or match ball rounds, reduced velocity ball rounds for use with silenced rifles, short range practice ball, plastic bulletted practice ball and various patented types of armour-piercing bullet.

DATA
Overall length 2.75in (69.85mm)
Case length 2.01in (51.05mm)
Rim diameter 0.470in (11.94mm)
Bullet weight 149 grains (9.65g)
Muzzle velocity
2,802ft/sec (854m/sec)
Muzzle energy
2,604ft/lb (3519J)
Velocity at 300 metres
2,106ft/sec (642m/sec)
Energy at 300 metres
1,471ft/lb (1,988J)

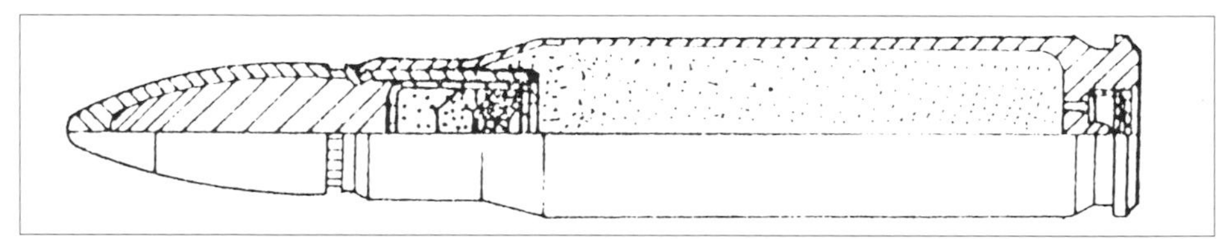

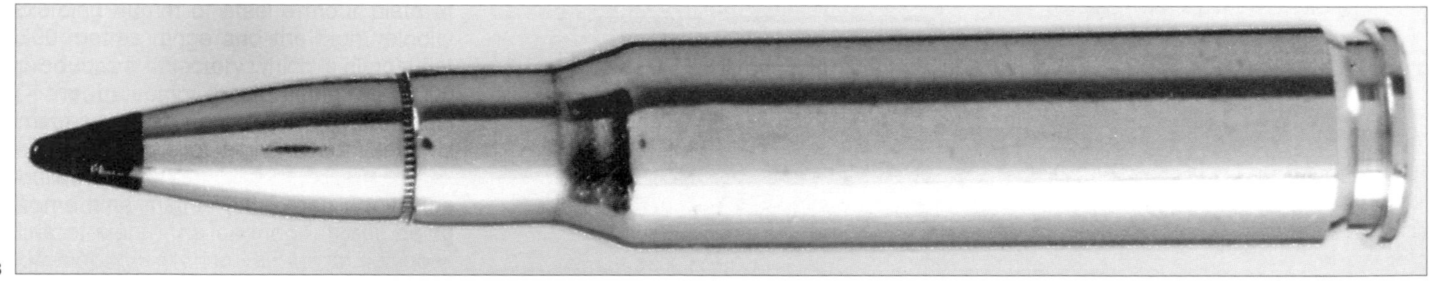

7.62 x 54R

Russia

The oldest military rifle cartridge surviving in active service, this dates from 1891 when it was introduced with the Russian Mosin-Nagant rifle. It has no particular advantage over any other cartridge of its period, having survived simply because it existed, there was ample production capacity and it did what was required. There was no reason to go to the expense and bother of changing it. It passed from being the standard rifle round when it was replaced by the 7.62 x 39mm M1943 cartridge, but it stayed on to provide the ammunition for several Russian machine guns and was also adopted for their Dragunov sniping rifle, since it had the desired range, accuracy and terminal effect. It was therefore adopted by other countries who used these weapons and is still widely manufactured.

The standard ball bullet has changed over the years; it was originally a round nosed, non-streamlined type weighing 215 grains, but a pointed bullet was adopted in 1909 and different weights were tried at various times until the final standard was a streamlined, pointed, steel cored compound bullet of 150 grains. There was also the usual range of armour-piercing, tracer, armour-piercing-incendiary, grenade launching blank and reduced load ball for silenced weapons. In addition, the Mosin-Nagant rifle was adopted by Finland, and the Finns then developed their own range of ammunition, differing slightly in detail from the Russian. Add on the minor variants made by Warsaw Pact countries, Serbia and one or two enterprising western manufacturers and there is a quite wide choice of ammunition available.

DATA
Overall length 3.019in (76.70mm)
Case length 2.110in (53.60mm)
Rim diameter 0.563in (14.30mm)
Bullet weight 185 grains (11.98g)
Muzzle velocity
2,638ft/sec (804m/sec)
Muzzle energy
2,822ft/lb (3,814J)
Velocity at 300 metres
1,988ft/sec (606m/sec)
Energy at 300 metres
1,603ft/lb (2,167J)

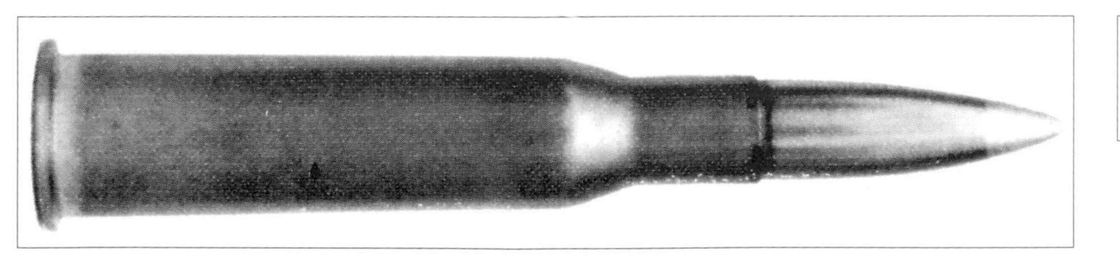

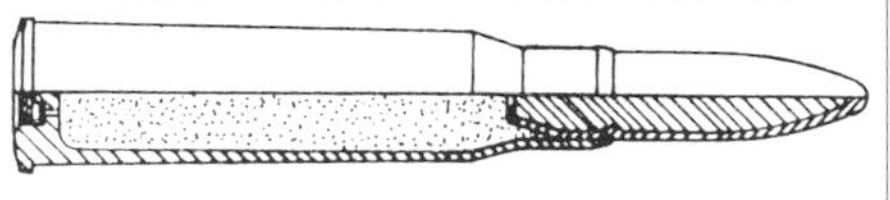

7.62mm Pistol Russia

This is the Russian version of the 7.63mm Mauser pistol cartridge. The Russians purchased quantities of Mauser pistols in the early 1900s and so did the Bolsheviks in the early 1920s, and having set up a manufacturing plant for the cartridges, they saw no reason to abandon it when they developed their own automatic pistols and chambered them for the same round. The barrels were made in 7.62mm calibre, because ample barrel making facilities existed in this calibre, and thus the cartridge adopted a new name. Then came the submachine guns, also in 7.62mm calibre and so on. It has now become a second line cartridge, since the old Tokarev pistol has been retired in favour of the Makarov, but there are still large numbers of weapons around which use the cartridge, not only in Russia, so that there is no likelihood of it vanishing for the foreseeable future.

The standard Russian bullet was the 'Ball Type P', jacketed with a lead core and weighing 86 grains. For submachine guns there was an armour-piercing-incendiary bullet and a tracer bullet, though this last appears to have rarely made an appearance. Manufacture of this cartridge is currently carried out in Russia, China and Serbia – all countries with stocks of elderly Russian weapons.

DATA
Overall length 1.360in (34.55mm)
Case length 0.989in (25.14mm)
Rim diameter 0.390in (9.91mm)
Bullet weight 86 grains (5.57g)
Muzzle velocity 1,492 ft/sec (455 m/sec)
Muzzle energy 426 ft/lbs (576 J)

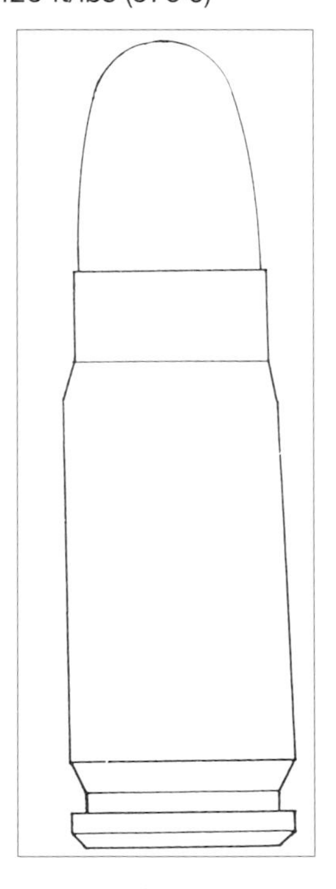

7.63 x 63mm Silent Pistol — Russia

The usual method of silencing a pistol (or any other firearm) is to direct the bullet and the emergent gases into a tube containing baffles and absorbent surfaces, so that the gas is slowed down before being allowed to escape to the open air. In this way it generates far less noise. This cartridge, however, demonstrates an entirely different way of silencing a pistol: keep the noise inside.

As can be seen, the case is abnormally long for a pistol cartridge and the bullet closely resembles, and probably is, the standard ball bullet used with the 7.62 x 39mm M1943 assault rifle cartridge. Inside the case is a propellant charge and in front of it a piston, with the base of the bullet seated on the tip of the piston rod. On firing, the propellant explodes and the piston is driven forward, pushing the bullet, until it is arrested by the case neck, which is firmly contained by the chamber of the pistol. The piston jams itself tight in the case and seals in all the explosion gas – and noise. The bullet is, in fact, catapulted out of the bore – the powder gases never touch it.

This doesn't break any velocity records, but it is not intended to. The object is to launch a pistol bullet with sufficient force to be lethal at short range and do it silently, and it achieves this perfectly.

DATA
Overall length 3.03in (77mm)
Case length 2.472in (62.8mm)
Rim diameter 0.500in (12.7mm)
Bullet weight 123 grains (8.0g)
Muzzle velocity
approx 655ft/sec (200m/sec)
Muzzle energy
approx 1,18ft/lb (160J)

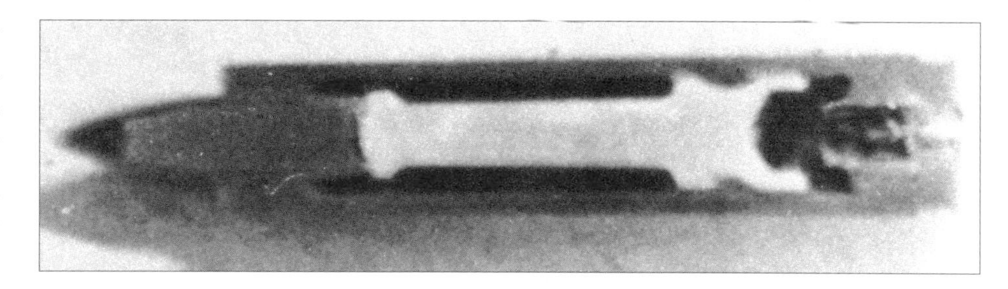

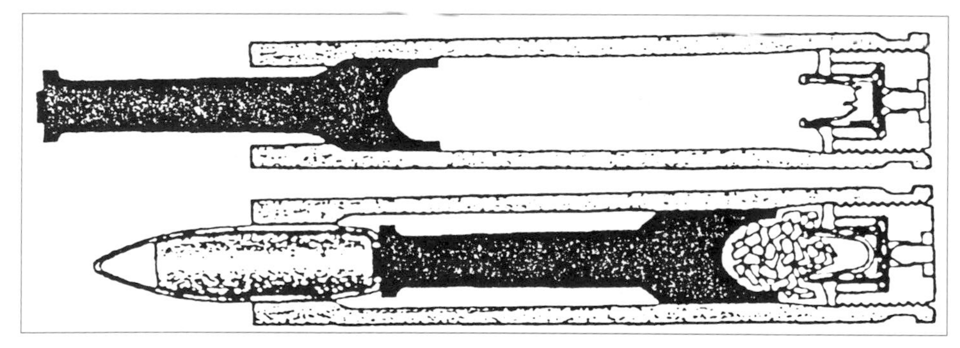

7.92 x 57mm Mauser

Germany

One of the most venerable cartridges in existence, this first appeared in 1888 with the German Commission rifle and continued, with periodic improvements, through a series of Mauser and other rifles as the standard German cartridge until 1945. It was also widely adopted in other countries, together with the parent rifles, and remains in military use to the present day, though largely with machine guns and sniping rifles. It is one of the old style, long range, powerful cartridges, delivering more punch at 300 metres than the 5.56mm round does at the muzzle. Just about every known type of bullet has been fired from this cartridge over the years, but the only military types in use today are ball, tracer and armour-piercing; there are numerous soft point and similar rounds available for hunting. The military round is currently manufactured in Germany, Portugal and Serbia; commercial (sporting) ammunition is manufactured in virtually every country except Great Britain.

DATA
Overall length 3.173in (80.60mm)
Case length 2.244in (57.00mm)
Rim diameter 0.472in (12.00mm)
Bullet weight 198 grains (12.85g)
Muzzle velocity 2,418ft/sec (737m/sec)
Muzzle energy 2,583ft/lb (3,490J)
Velocity at 300 metres
1,818ft/sec (554m/sec)
Energy at 300 metres
1,459ft/lb (1972J)

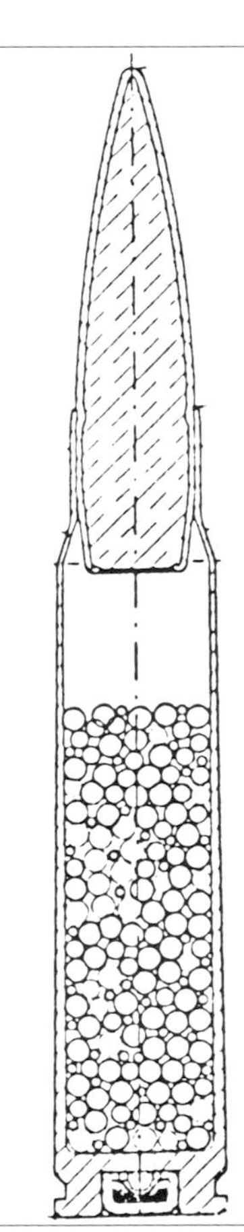

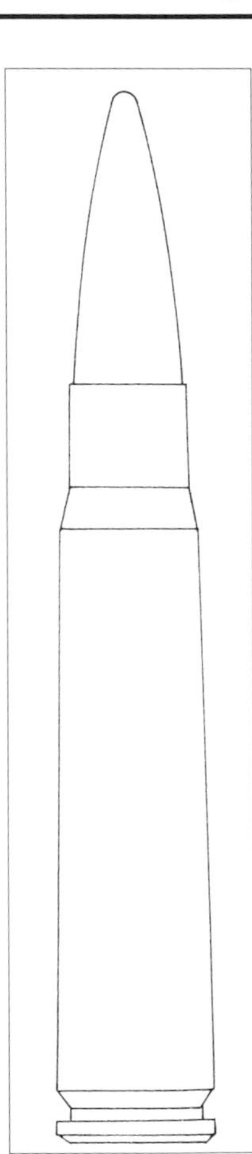

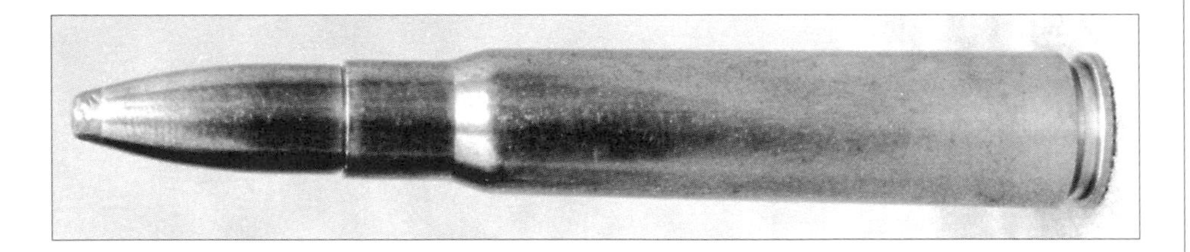

.338 Lapua Magnum

Finland

This was developed by Lapua of Finland as a long range target and hunting round, the cartridge case having been adapted from the existing .416 Rigby sporting cartridge by reducing the neck diameter to suit the bullet. It soon acquired a reputation for superb accuracy in suitable rifles and, carrying a jacketed ball bullet, has been suggested by several experts as a sniping cartridge. At the present time only a few sniping rifles have been chambered for it, but the results are impressive and it can only be a matter of time before it becomes accepted as a military round. The sniping bullet is a jacketed compound type.

DATA
Overall length 3.602in (91.50mm)
Case length 2.724in (69.20mm)
Rim diameter 0.587in (14.91mm)
Bullet weight 250 grains (16.20g)
Muzzle velocity 3,000ft/sec (914m/sec)
Muzzle energy 5,007ft/lbs (6766J)
Velocity at 400 metres
2,339ft/sec (713m/sec)
Energy at 400 metres
3,054 ft/lbs (4,127J)

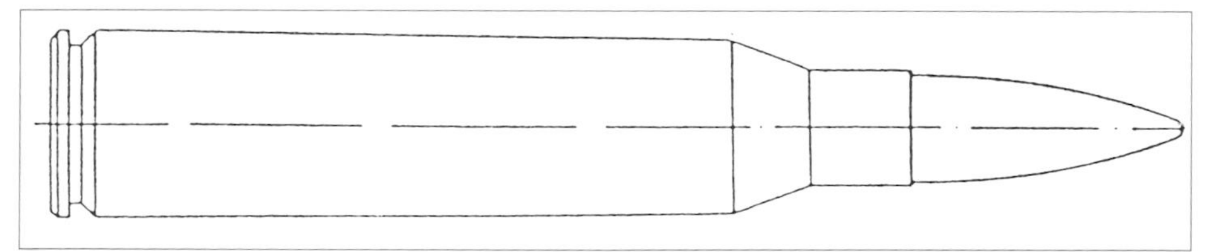

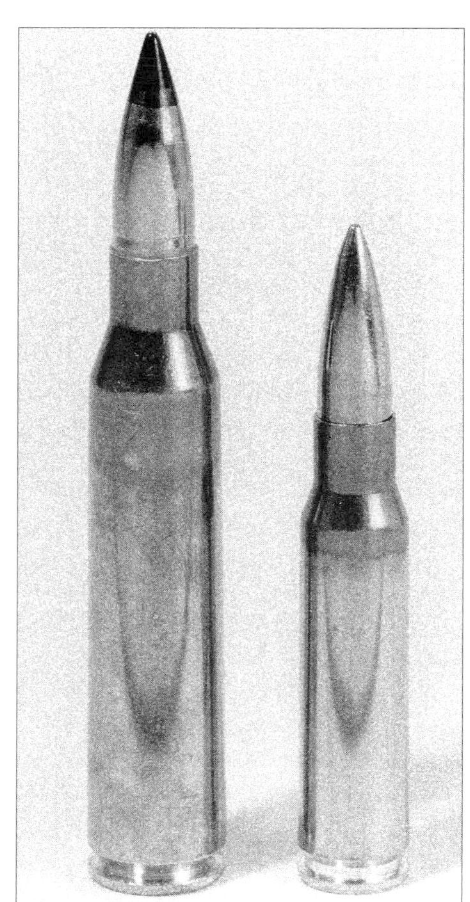

(Right) The .338 Lapua Magnum compared with a 7.62 x 51mm NATO round.

9mm Makarov **Russia**

This cartridge takes its name from its appearance in Soviet service with the Makarov pistol in 1951, though it was not generally known in the West until the early 1960s. The development history is not known, but it was probably influenced by pre-war work in Germany on the 'Ultra' series of cartridges, since the object in view appears to be the same – the maximum power from a 9mm round fired in a blowback pistol. It falls more or less midway between the 9mm Short and 9mm Parabellum rounds for performance. The metric designation is 9 x 18mm and this can cause confusion with the 9 x 18mm police cartridge, of similar size and shape, but not interchangeable with the Makarov; the police round was developed in Germany but has not been the success that was envisaged, and may well disappear entirely in a few years. The Makarov, though, will be around for some time to come, since it was widely distributed throughout the Warsaw Pact countries and to other countries supplied with Soviet weapons. Ammunition is currently made in Bulgaria, China, Czech Republic, Hungary, Italy, Poland, Romania, Russia, Serbia and the USA.

DATA
Overall length 0.975in (24.79mm)
Case length 0.708in (17.98mm)
Rim diameter 0.393in (9.98mm)
Bullet weight 94 grains (6.10g)
Muzzle velocity 1,017ft/sec (310m/sec)
Muzzle energy 257ft/lb (348J)

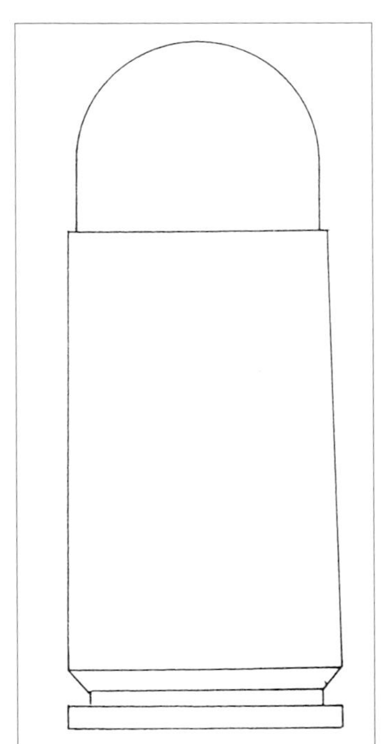

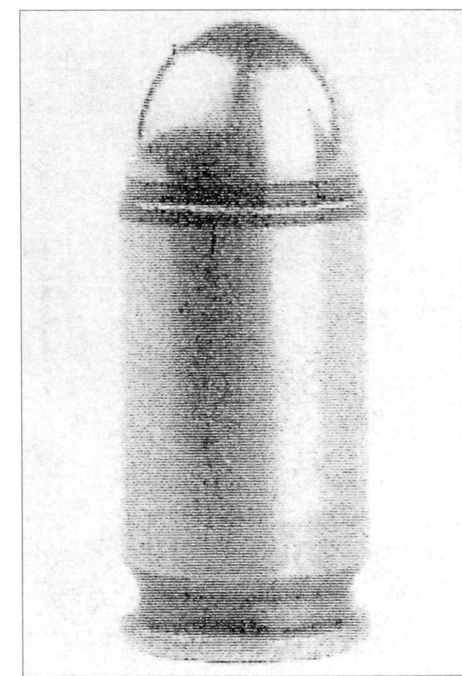

9mm Parabellum

Virtually the standard military pistol and submachine gun cartridge of most of the world, this dates back to 1902 when Georg Luger improved the stopping power of his automatic pistol by opening up the case mouth of the then standard 7.65mm cartridge and putting a 9mm bullet in it. Over the years it has seen a few minor improvements and a wide variety of bullets, but the standard military round is now a jacketed lead-cored ball. There are also a number of short range expanding bullets intended for use by police forces, plastic bullets for practice, tracer rounds and subsonic rounds for silenced weapons. There are also special rounds for submachine guns, with heavier charges and armour-piercing cores. It would be pointless to list the countries manufacturing this cartridge; any country with an ammunition industry will have the 9mm Parabellum in steady production.

DATA (British Mark 2Z)
Overall length 1.152in (29.28mm)
Case length 0.762in (19.35mm)
Rim diameter 0.391in (9.94mm)
Bullet weight 115 grains (7.45g)
Muzzle velocity 1,300ft/sec (396m/sec)
Muzzle energy 431ft/lb (583J)

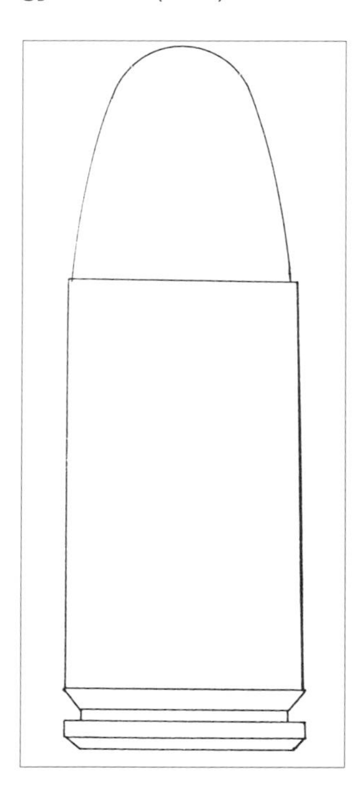

.357 SIG

This round was developed by Sigarms of the USA in an attempt to persuade users of the .357 revolver to adopt an automatic pistol.

It is more or less the .40 Auto cartridge case necked down to take a standard .357 Magnum revolver bullet, which allows for a case with sufficient volume to carry a charge which will deliver similar ballistics to the revolver. Whether it is having the desired effect upon revolver owners is not yet apparent, but it is sufficiently compact and powerful to make it attractive to military and security forces. Manufacture is currently confined to two companies in the USA.

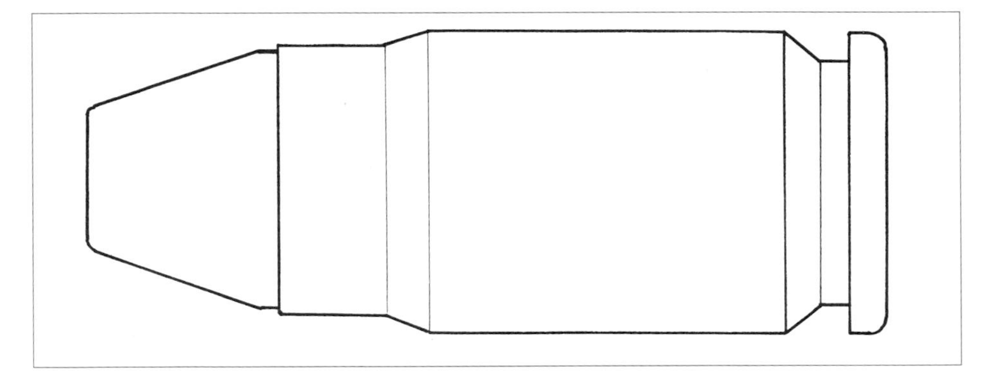

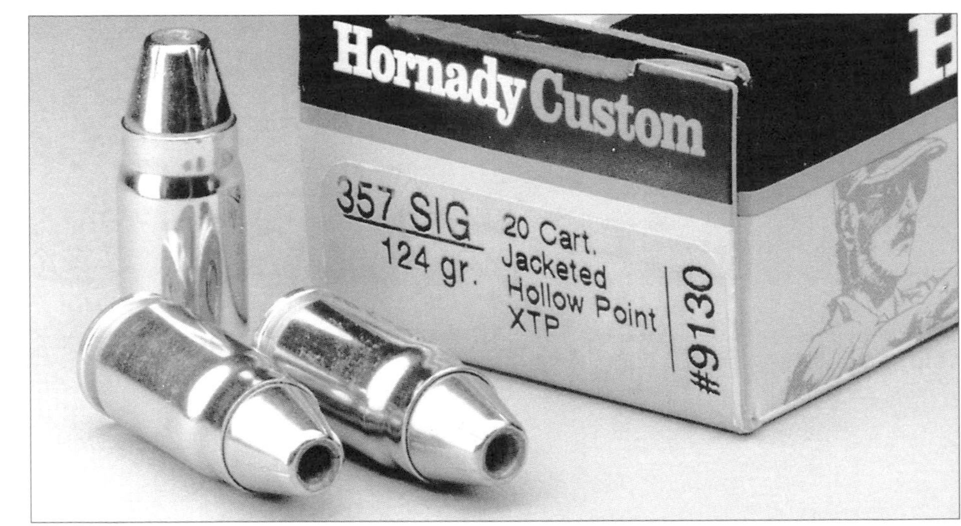

DATA
Overall length 1.120in (28.45mm)
Case length 0.860in (21.84mm)
Rim diameter 0.419in (10.64mm)
Bullet weight 125 grains (8.10g)
Muzzle velocity 1,352ft/sec (412m/sec)
Muzzle energy 508ft/lb (686J)

.38 Special

Another venerable round which retains its popularity in sporting circles and, due to its good ballistics and consistent performance, appears in several military inventories. It is particularly popular with aircrew and military police personal weapons. It first appeared in 1900 as a potential military cartridge, but the US Army were intent upon a .45 calibre and the round became a commercial success. The long case prevents it being chambered in revolvers too weak to stand the loading and it can be found with every type and weight of bullet possible. The current military standard bullet is a lead-cored jacketed type. Like the 9mm Parabellum, its popularity is such that it is made in virtually every country which has an ammunition industry.

DATA (US M41 Ball)
Overall length 1.550in (39.37mm)
Case length 1.159in (29.45mm)
Rim diameter 0.433in (11.00mm)
Bullet weight 132 grains (8.55g)
Muzzle velocity 948ft/sec (289m/sec)
Muzzle energy 264ft/lb (357J)

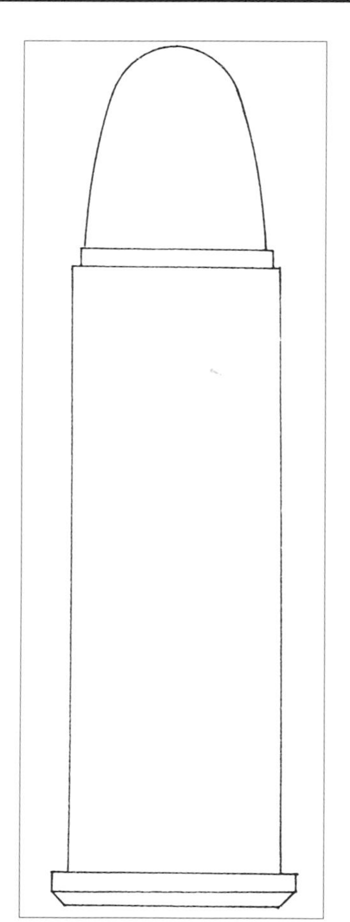

.40 Smith & Wesson

This appeared in the late 1980s as Smith & Wesson's response to the 10mm Auto cartridge. It has the advantage of being somewhat shorter than the 10mm Auto, thus a pistol frame originally designed for the 9mm Parabellum cartridge can be fairly easily modified to build a .40 pistol. The longer case of the 10mm round demands a complete new frame. The calibre has been adopted by several of the major weapon manufacturers, and pistols and submachine guns have been made. It appears to have gained a larger following than the 10mm Auto, though it has not yet been formally adopted by any military force. The cartridge case has a rebated rim, smaller than the case base, which is another point which makes the conversion of a 9mm weapon design easier. For police use the usual bullet is a semi-jacketed soft point, but a full metal jacket bullet has been developed for possible military adoption. Ammunition is currently made in Israel, Italy, South Korea and by several firms in the USA.

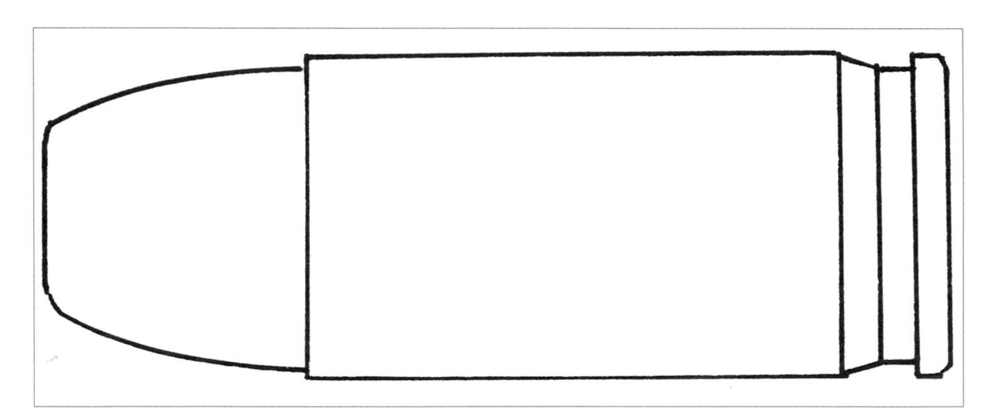

DATA

Overall length 1.130in (28.70mm)
Case length 0.850 (21.60mm)
Rim diameter 0.424 (10.77mm)
Bullet weight 178 grains (11.55g)
Muzzle velocity 935ft/sec (285m/sec)
Muzzle energy 347ft/lb (469J)

10mm Auto

The 10mm Auto is an idea which has been around for most of the 20th century without ever achieving any degree of success. This particular version was developed together with the 10mm 'Bren Ten' pistol in 1980–83, the object being to produce a pistol with recoil comparable to the standard .45 ACP, but with superior power. The Bren Ten failed, but the Norma company had begun producing ammunition and various other makers looked at the idea and began producing pistols. Colt appeared with their 10mm Delta, followed by Smith & Wesson and the cartridge finally achieved respectability when it was adopted by the Federal Bureau of Investigation as their standard service round. This led to greater interest and more weapons in this calibre, both pistols and submachine guns, have since appeared. Various types of bullet are available, but the military standard is the jacketed lead-cored type. Ammunition is currently being made in Brazil, Italy, South Korea, Sweden and the USA.

DATA
Overall length 1.252in (31.78mm)
Case length 0.984in (25.00mm)
Rim diameter 0.422in (10.72mm)
Bullet weight 170 grains (11.00g)
Muzzle velocity 1,102ft/sec (336m/sec)
Muzzle energy 637ft/lb (861J)

.45 ACP

From time to time various contenders have appeared, intent upon removing the .45 ACP from its position as the favourite American pistol cartridge, but they inevitably fail and the Colt M1911 looks like making its century with little difficulty. Although formally ousted by the 9mm Parabellum as the US service pistol round, it was noticeable that when Special Forces demanded their own side arm, .45 was the specified calibre.

The military loading of this cartridge has remained the same since the time of its introduction, apart from alterations in the nature of the powder from time to time. The bullet has always been a jacketed, lead-cored, 230 grain article with impressive stopping power. Other types, of every description and a wide range of weights, have been offered commercially over the years and since the pistol has been sold world wide, attracting a number of imitators, the ammunition is made in almost every country outside the Soviet influenced parts of the world.

DATA (US M1911 Ball)
Overall length 1.267in (32.19mm)
Case length 0.897in (22.79mm)
Rim diameter 0.467in (11.86)
Bullet weight 230 grains (14.90g)
Muzzle velocity 804ft/sec (245m/sec)
Muzzle energy 330ft/lb (446J)

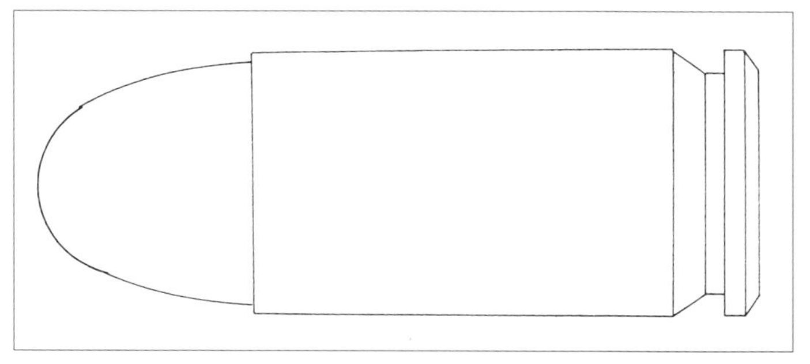

.50 Browning

The history of this remarkable cartridge starts in 1918 when General Pershing called for a heavy anti-aircraft machine gun. The war ended before much was done, but in 1919 an enlarged Browning machine gun had been developed and this cartridge was designed by simply scaling up the existing .30 Model 1906 rifle cartridge. It came into its own during World War Two and after 1945 the gun and cartridge were adopted by almost every nation outside the orbit of the Soviet Union. By remarkable applications of technology the ammunition is still holding its place, even though its demise has been forecast for some years; every attempt to supplant it by a fresh design of machine gun and cartridge has failed, largely due to the vast numbers of guns in use, which would be expensive to replace. The adoption of the cartridge for use in heavy anti-matériel rifles can only extend its life even further.

DATA
Overall length 5.425in (137.80mm)
Case length 3.90in (99.10mm)
Rim diameter 0.800in (20.30mm)
Bullet weight 662 grains (42.90g)
Muzzle velocity 2,910ft/sec (887m/sec)
Muzzle energy 12,488ft/lb (16,876J)
Velocity at 300 metres
2,182ft/sec (665m/sec)
Energy at 300 metres
7,012ft/lb (9,476J)

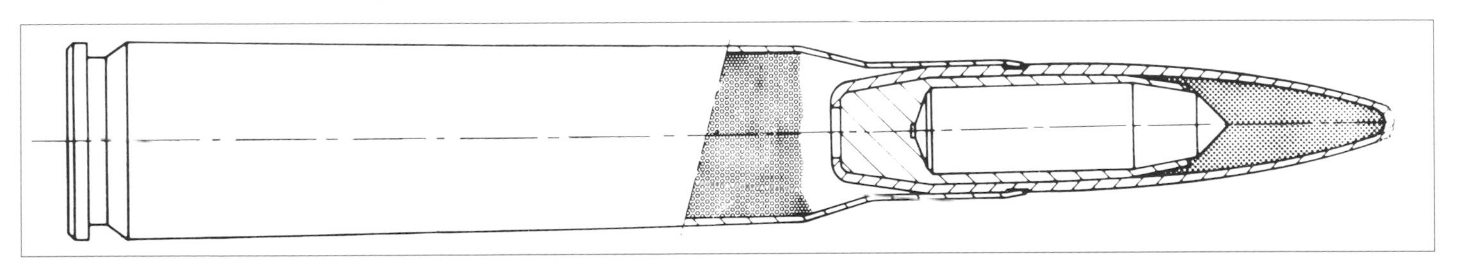

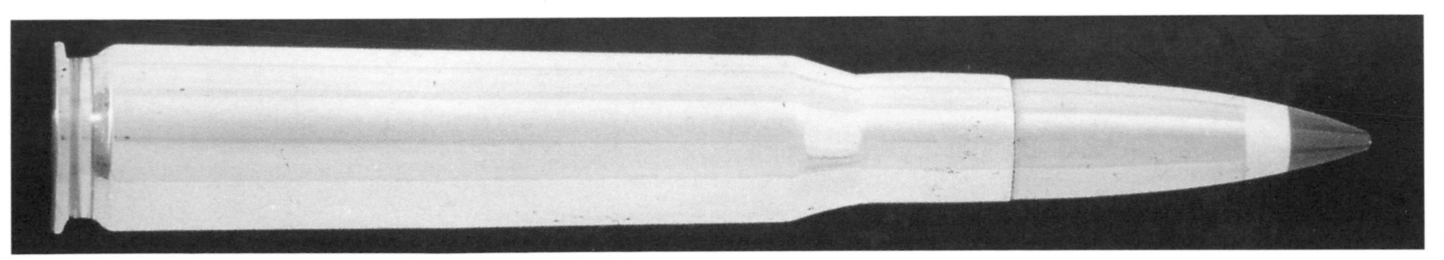

12.7 x 107mm Russia

This might be said to be the Soviet answer to the .50 Browning, being of similar shape, size and performance. It was developed in the late 1920s and is said to have benefited from the German 13mm Tank und Flieger cartridge under development in 1918 for much of its inspiration, though it might equally well have been based upon the Browning round. It was first used with the Degtyarev DK heavy machine gun in the mid 1930s and then the modified DShK guns, and has remained in use ever since. It has been widely distributed to the many countries under Soviet influence in the post-war years and made under licence in China and the Middle East. With the American pioneering in heavy rifles using the .50 Browning, it is hardly surprising that similar designs adapted to this cartridge have appeared in countries in which it is a common round. As far as performance in rifles is concerned there is little solid information, but it should be on a par, both for accuracy and destructive power, with the Browning round.

DATA
Overall length 5.780in (146.8mm)
Case length 4.169in (105.9mm)
Rim diameter 0.850in (21.6mm)
Bullet weight 745 grains (48.28g)
Muzzle velocity 2,756ft/sec (840m/s)
Muzzle energy 11,522ft/lb (15,570J)
Velocity at 300 metres
2,067ft/sec (630m/sec)
Energy at 300 metres
7,083 ft/lb (9572J)

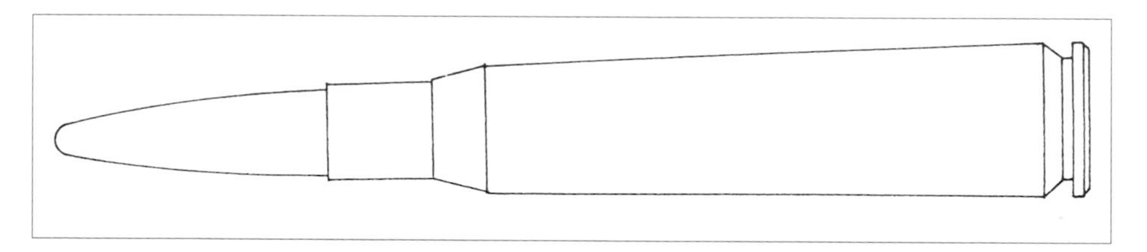

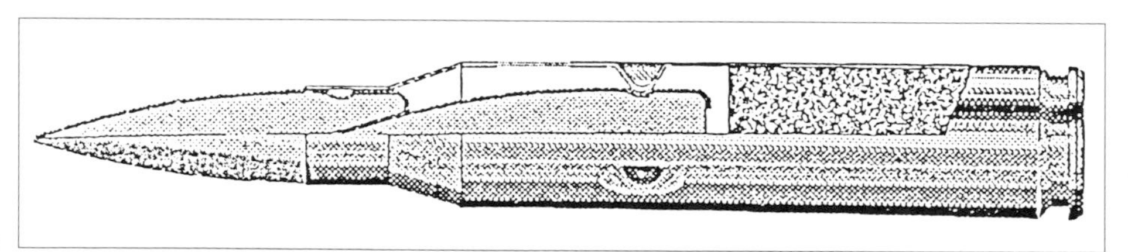

(Left) The Duplex principle applied to the Soviet 12.7mm cartridge.

14.5 x 114mm Russia

Once the 12.7mm Soviet round had appeared in a rifle it was only a matter of time before this 14.5mm cartridge was tried in the same role. This monster was devised as an anti- tank rifle cartridge and it certainly fuelled the most potent of all the World War Two anti-tank rifles, the Russians using them long after everyone else had abandoned them. After the war its performance was too good to lose and the KPV heavy machine gun was built around it, proving to be a useful air defence weapon. The steel-cored armour piercing bullet can penetrate 28mm of armour at 300 metres range and the figures below indicate that it will retain quite adequate power to deal with light armour well down range. There are also armour-piercing/incendiary bullets and a high explosive incendiary which can go through duralumin plate at 1,500 metres and ignite a fuel tank behind it. However, a 63 gram bullet leaving the barrel at 3,200 feet per second does produce a hefty 'kick'.

DATA
Overall length 6.142in (156mm)
Case length 4.50in (114.3mm)
Rim diameter 1.059in (26.9mm)
Bullet weight 980 grains (63.44g)
Muzzle velocity 3,200ft/sec (976m/sec)
Muzzle energy 22,360ft/lb (30,215J)
Velocity at 300 metres
2,402ft/sec (732m/sec)
Energy at 300 metres
12,565ft/lb (16,979J)

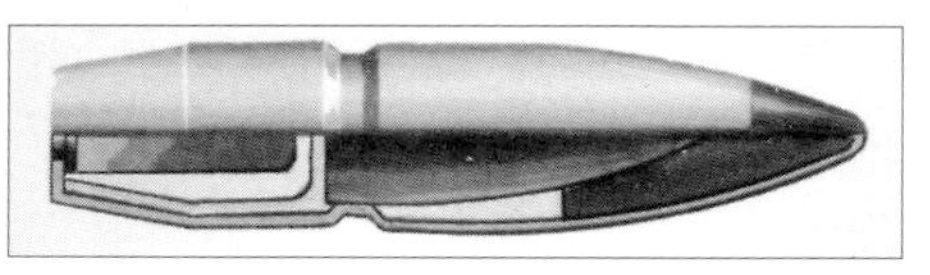

(Above) The 14.5mm AP-Incendiary-Tracer bullet.

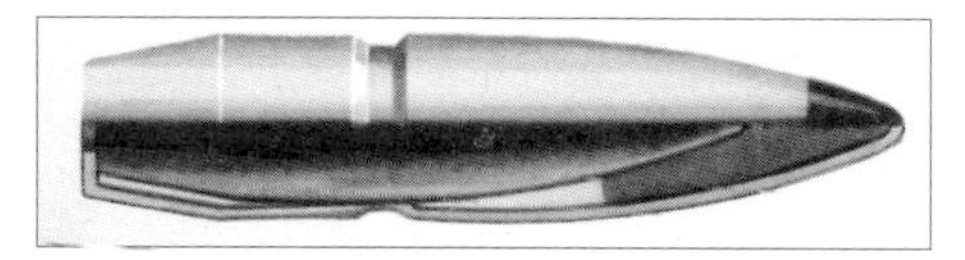

(Above) The 14.5mm AP-Incendiary bullet.

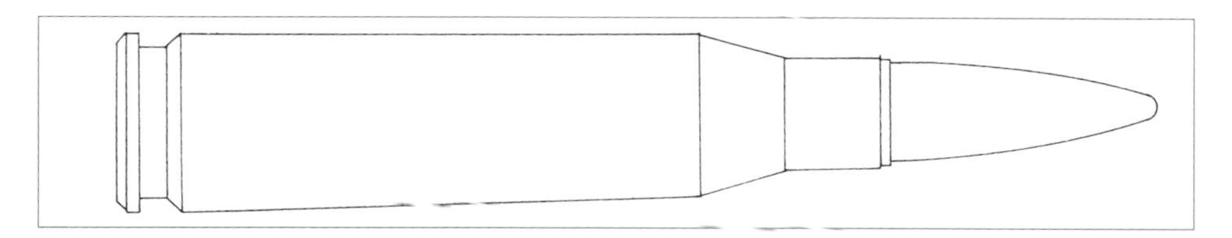

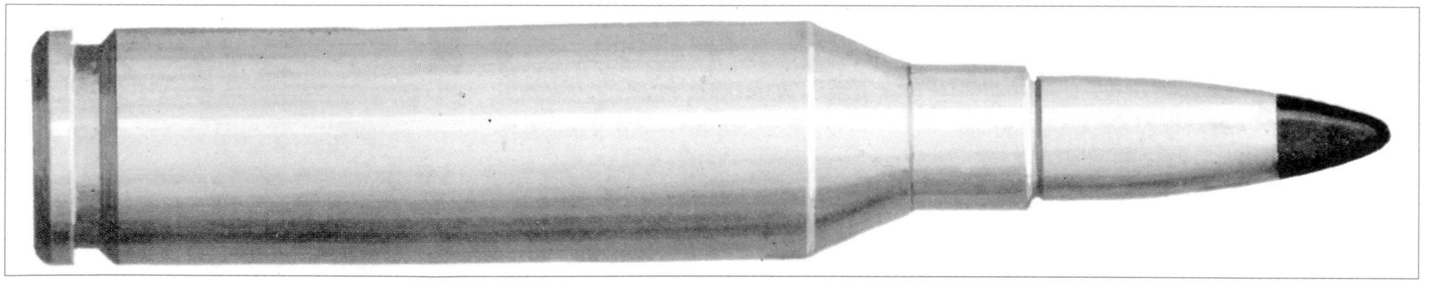

15.2 Steyr AMR

Austria

This is an unusual cartridge for an unusual weapon. The Steyr Anti-Matériel Rifle (AMR) is a heavy two-man weapon designed to deal out severe damage to high-tech targets at long range with superb accuracy, and the ammunition is therefore somewhat different. The cartridge case is of compound construction, part metal and part synthetic material, though it is of conventional bottle necked form. The projectile is a fin stabilised tungsten dart of 5.5mm diameter, held in a plastic discarding sabot supported on a pusher piston. The cartridge has a long primer which ensures adequate and thorough ignition of the charge and the explosion drives the piston which propels the dart up the bore. At the muzzle the sabot segments are thrown off, air drag causes the piston to fall clear and the dart goes on to the target. It has shown itself capable of piercing 40mm of steel armour plate at 800 metres range and the high velocity produces a trajectory which is almost flat – the trajectory peak, firing at 1,000 metres range, is only 80cm (31.4 inches) above the line of sight; a conventional bullet, at the same range, has a vertex some nine metres (29.5 feet) above the line of sight. The cartridge is still being developed and the rifle has not yet been adopted by any military force, though several are evaluating it.

DATA

Overall length 8.150in (207mm)
Case length 6.693in (170mm)
Rim diameter 1.023in (26.00mm)
Bullet weight 540 grains (35.00g)
Muzzle velocity
4,757ft/sec (1,450m/sec)
Muzzle energy
27,225ft/lb (36,792J)

(Below) The component parts of the Steyr 15.2mm anti-matériel round.

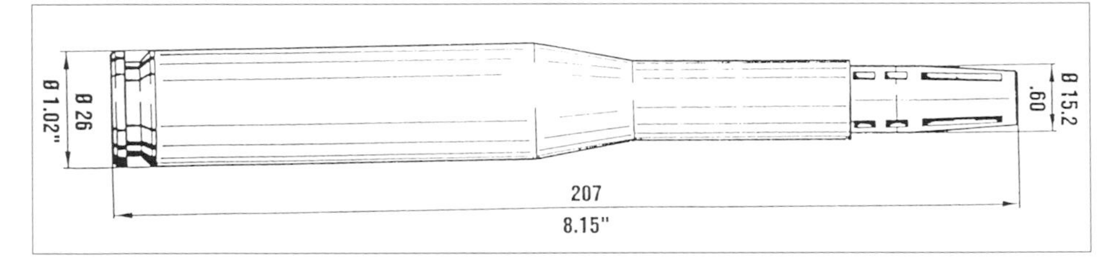

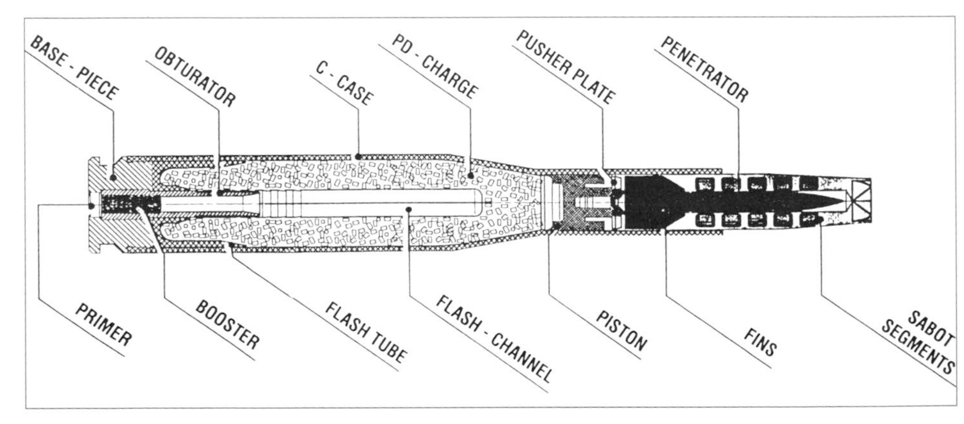

Small Arms and Cannon Ammunition: Cannon

Insofar as it consists of a projectile firmly fixed into a cartridge case, cannon ammunition resembles small arms ammunition but is bigger. However, the significant difference lies in the projectile which, due to its greater size, allows the designer greater freedom in design and allows the development of high explosive shells, special piercing projectiles, incendiary shells and various other projectiles. Instead of treating the projectile as an overgrown bullet and forcing it to squeeze itself into the rifling, it is possible to treat it as a scaled down artillery projectile, giving it a driving band to take the rifling and a separate fuze to provide better control over the target effect.

The automatic cannon has a curious history. It really began as a development of the machine gun in the 1890s when both Maxim and Hotchkiss produced enlarged examples of their guns to fire 37mm explosive shells. This calibre was determined by the St Petersburg Convention which laid down that explosive projectiles weighing less than 400 grams would be prohibited in war and part of the protocol which outlawed 'explosive' small arms bullets. The smallest practical high explosive shell which could be made to weigh more than 400 grams proved to be 37mm in calibre and that is how that peculiar calibre came into existence. Both the Maxim and Hotchkiss were bought by navies to use as short range weapons against torpedo boats. The Maxim achieved fame in the Boer War as the 'Pom-Pom', being used as a light field weapon by the Boers. Although it proved something of an irritant to the British, it was never a serious threat and no army ever managed to work out precisely what use such a weapon had in open warfare. The shell was too light to do real damage to fieldworks, it did not produce a big enough lethal area to make it cost effective against personnel and the range was insufficient if treated as artillery and too great if treated as an infantry machine gun. It found a niche in 1914–18 as a light anti-aircraft gun, but one tends to read between the lines of contemporary reports and reach the conclusion that it was placed in that role because nothing better could be found for it.

Later in the same war two inventors appeared who saw a totally different role to be filled: that of aerial warfare. Contemporary aircraft were all armed with rifle calibre machine guns and by 1917 it was becoming apparent that such weapons were marginally effective against the bigger bombers and even against Zeppelins. The brothers Conders, employees of the Stahlwerk Becker of Reinickendorf, Germany, developed a form of heavy machine gun in 19mm calibre, firing a small high explosive shell. This was well below the 400 gram limit, but was rationalised as being excusable since it was designed to be fired at aircraft, not men. Moreover, there was a war on, a situation in which such rationalisations tend to be acceptable. Some guns were installed on Gotha bombers, about 130 were issued to anti-aircraft troops and in all some 392 were seized and destroyed by the Allied Disarmament Commission after the war. The Becker company was forbidden to manufacture firearms, so it sold its patents to a Swiss firm, the Seebach Engineering Company. They made a few changes, altered the calibre to 20mm, put the gun on wheels and began promoting it as an anti-tank and infantry support gun. They had no success and in 1924 went into liquidation. The factory, tools, patents and drawings were bought up by another Swiss firm, the Oerlikon Machine Tool Company of Zurich. Oerlikon promoted the Oerlikon cannon as both an anti-aircraft and aircraft weapon with great success and it has been at the forefront of cannon and cannon ammunition design, ever since. Once the idea had been accepted, other designs appeared, Hispano Suiza being probably the best known, and towards the end of World War Two, when aircraft were becoming more resistant to attack, the Germans, using a Mauser design, increased the calibre from 20mm to 30mm.

30mm is the upper limit of cannon calibres; once past this figure the weapon becomes classed as artillery. Therefore the gap between 20 and 30mm has been fairly exhaustively explored and three intermediate calibres, 23, 25 and 27mm, have proved workable.

The automatic cannon today has three applications: as an aircraft weapon, as an anti-aircraft weapon, and as armament for light armoured combat vehicles, or infantry fighting vehicles. In the two former roles the object is firstly to attack other aircraft and secondly to attack ground targets; in the third role the prime object is to attack other similarly armoured vehicles. As a result, cannon ammunition tends to fall into two groups, that for the attack of relatively 'soft' targets and that for the attack of distinctly harder ones.

For soft targets the primary projectile is the high explosive/incendiary (HE-I) shell. The shell body is of high strength steel so as to be thin and carry a useful charge of explosive. The explosive is an aluminized Hexogen/TNT compound which produces an incendiary effect as it detonates, without needing the inclusion of any special incendiary mixture. An impact fuze in the nose detonates the shell on impact. An alternative, designed to deal with lightly armoured aircraft and with lightly armoured ground targets, is the armour-piercing HE-I shell; this differs in having a sharply pointed and thick walled body, with a hard tip having a somewhat smaller filling of explosive and a base fuze in the bottom of the shell. This can thus penetrate light protection and then detonate behind it.

For hard targets the armour piercing HE-I shell can be used, but more and more designers have turned to adaptations of heavier tank ammunition and have produced various types of discarding sabot projectile. This consists of an exceptionally hard core of tungsten carbine or depleted uranium, carried in a light alloy sheath. This sheath is so designed that under the sudden acceleration of being fired it splits into segments which are retained in place by the fact of being inside the gun barrel. They therefore grip the core and transmit spin to it from the gun's rifling. On leaving the muzzle this spin causes the individual parts of the sheath (or sabot) to fly off and leave the core to go on to the target. Since the complete projectile, as fired up the bore, is lighter than a conventional shell, it reaches a very high velocity. As the extraneous sabot is thrown clear at the muzzle, what is left is a small calibre, very hard, core travelling at high speed and hence having considerable penetrative ability. As an example, a 25mm armour-piercing DS core will penetrate 25mm of armour at 2,000 metres range, striking at an angle of 30^{0} to the target plate.

An improvement on this can be achieved by adopting a fin stabilised sub-projectile resembling a dart. This can be made longer than a spin stabilised sub-projectile therefore heavier for a given calibre, which means more kinetic energy and more penetration. The same 25mm gun firing an armour-piercing, FSDS round can defeat 36mm of armour at 1,000 yards range or 31mm at 2,000 yards, again striking at 30° to the target plate.

The most recent innovation in cannon calibres is the development of Cased Telescoped Ammunition. In simple terms this means placing the projectile inside the cartridge case, surrounded by propellant. This makes for a very compact cartridge which in turn allows the design of the gun to be more compact and it has been proved that it is possible to make a 45mm calibre gun using CTA ammunition which is very little bigger than a 30mm gun, but more powerful than a 35mm weapon. To achieve this, though, has meant some ingenious thinking and some unconventional design.

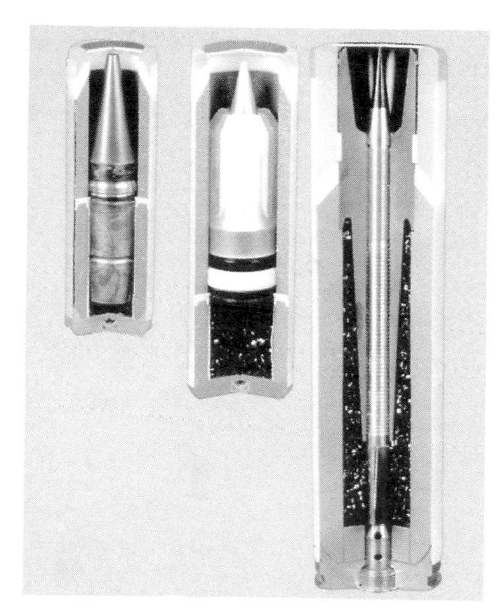

Examples of cased telescoped ammunition (CTA) rounds.

The major problem is that of placing the projectile into the bore of the weapon before the propelling charge generates the necessary propulsive gas. If one merely puts the bullet into the case, surrounded by propellant and fires it, much of the explosion gas will rush up the bore before the bullet gets into the rifling. So the projectile has a plastic sleeve around the front end to centre it in the case and also to close the case and keep the propellant inside. When the cap fires, a tiny charge is also fired and this pushes the bullet forward and jams the plastic sleeve tight against the rear face of the barrel, thus providing a gas seal and allowing the projectile to continue forward into the barrel and engage with the rifling. Then the main propelling charge explodes and drives the projectile up the bore in the usual manner.

There have been proposals for CTA in rifle and machine gun calibres and a .50 inch CTA machine gun has been built and fired. But there seems very little point in applying this technology in small arms calibres; like so many other proposed improvements in small arms, the expense would be enormous and the advantages relatively minor. In cannon calibres, though, there is a definite gain in power for a given size of weapon. Moreover, an entirely new operating system has been developed which takes advantage of the compact design of the round of ammunition and, by removal of the conventional reciprocating bolt or breech block, results is a more compact gun. The gun chamber is arranged to pivot, so that the two ends are exposed on opposite sides of the gun. The round is fed into the rear end and the chamber pivots back to lie between the barrel and the fixed breech piece. The round is fired and the chamber pivots once more; a fresh round is pushed into the rear of the chamber, pushing the empty case out on the other side of the gun, and the chamber swings closed. For this reason the base of the cartridge case is convex, though the designers appear to have catered for possible variations by providing the case with an extraction groove.

Current thinking is for the CTA 45mm gun to be possibly adopted for use in forthcoming infantry fighting vehicles. Ammunition developed for it includes an HE shell, an armour-piercing FSDS sabot projectile, target practice rounds for both the foregoing and a special 'frangible armour piercing DS' which appears to combine armour piercing and incendiary effect.

Current standard cannon rounds are listed below. Details given are for the normal HE-I projectile, but other types of ammunition available are indicated as follows:

HE High explosive; HE-I High explosive, incendiary; AP armour-piercing; AP-I armour-piercing, incendiary; AP-HC armour-piercing, Hard Core (Tungsten); TP Target practice; APDS armour piercing discarding sabot; FAPDS Frangible APDS; APFSDS Armour piercing fin stabilised discarding sabot; AMDS Anti Matériel Discarding Sabot (like armour piercing APDS, but for softer targets such as anti-ship missiles); MP Multi-purpose (Raufoss pattern). The addition of T indicates a tracer round.

20 x 82mm MG151

This cartridge was developed by Mauser of Germany in the 1930s as a possible anti-tank round, but it eventually found its niche in aircraft and anti-aircraft weapons in both Germany and Japan. After 1945 the cannon and ammunition were adopted in France and were sold to numerous countries. Manufacture of the ammunition has continued in France and in the late 1980s the South Africans adopted it for a cannon of their own design. It has also been put forward as a suitable cartridge for use in anti-matériel rifles.

DATA
Round length 5.787in (147mm)
Case length 3.216in (81.7mm)
Rim diameter 0.988in (25.1mm)
Projectile diameter 0.783in (19.9mm)
Projectile weight 3.88oz (110g)
Muzzle velocity 2,362ft/sec (720m/sec)
Muzzle energy 21,090ft/lbs (28.5kJ)
Types of projectile available
AP-I, HE-I-T, AP-I-T, TP-T

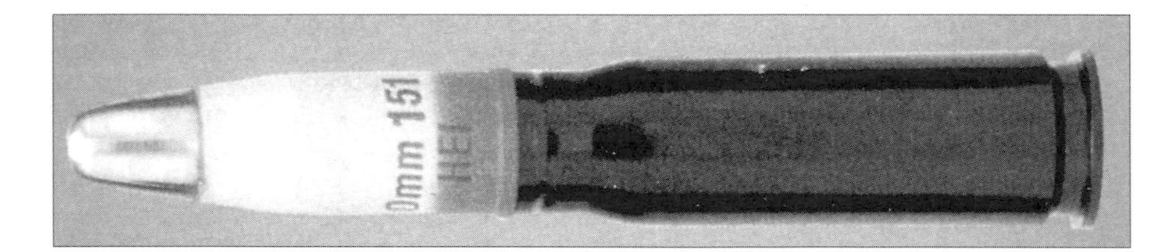

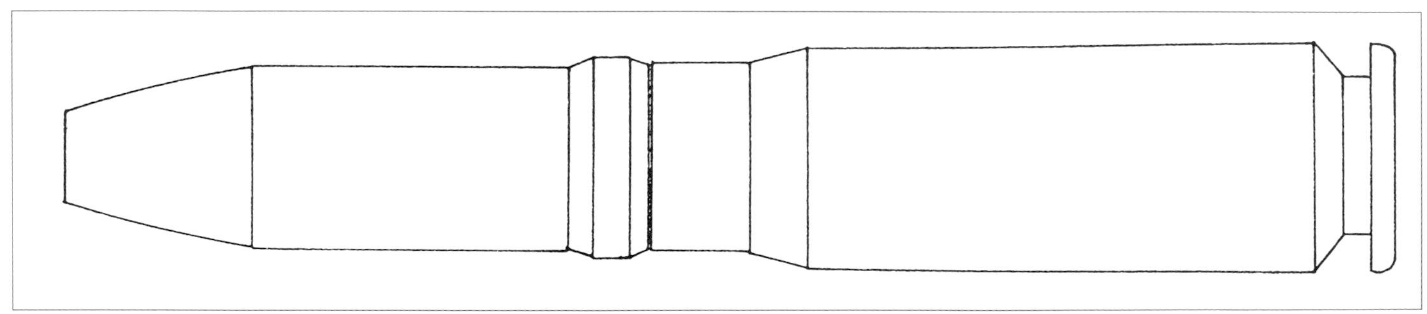

20 x 102mm US M39

USA

After wartime experience with the Oerlikon and Hispano-Suiza aircraft cannons, the US Air Force decided to set about designing one of its own, based on the Mauser 312 revolver gun. This entered service as the M39 cannon and the round of ammunition which was developed for it took the same number. It was later adopted for the M61 and M168 Vulcan Gatling guns and for some naval cannon. The guns and their ammunition then came into NATO use via the USA and have also appeared in other countries supplied with American aircraft. Manufacture currently takes place in Brazil, Canada, France, Germany, Greece, Italy, the Netherlands, Norway, Pakistan, Singapore, South Korea, Spain, Turkey and the USA.

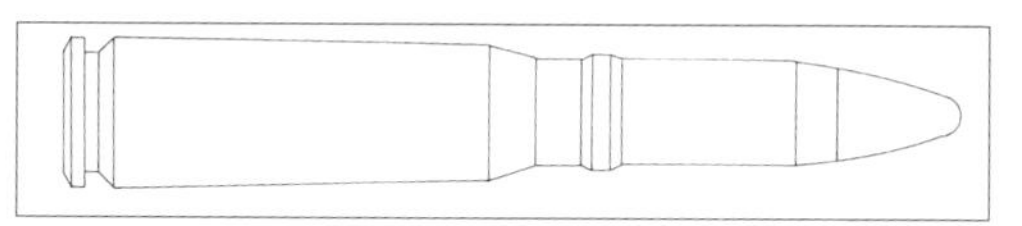

DATA

Round length 6.61in (168.0mm)
Case length 4.015in (102.0mm)
Rim diameter 1.165in (29.6mm)
Projectile diameter 0.783in (19.9mm)
Projectile weight 3.56oz (101g)
Muzzle velocity
3,380ft/sec (1,030m/sec)
Muzzle energy
39,590ft/lb (53.5kJ)
Types of projectile available
HE-I, HE-I-T, AP-I, AP-T, TP, TP-T, APDS-T, MP-T

20 x 110mm HS404

This takes its name from the Hispano-Suiza 404 cannon with which it was first introduced and which armed the French Dewoitine D-501 fighter aircraft in the mid 1930s. The weapon was subsequently used by Britain and the USA during World War Two and in post war years was adopted in many countries. Ammunition is currently manufactured in Argentina, Brazil, Egypt, Finland, France, Italy, Norway, Spain, Switzerland and Serbia.

DATA
Round length 7.244in (184.0mm)
Case length 4.335in (110.1mm)
Rim diameter 0.965in (24.5mm)
Projectile diameter 0.783in (19.90mm)
Projectile weight 4.30oz (122g)
Muzzle velocity 2,770ft/sec (844m/sec)
Muzzle energy 32,190ft/lb (43.5kJ)
Types of projectile available
HE-T, HE-I, HE-I-T, AP-I-T, AP-T, TP-T, MP-T

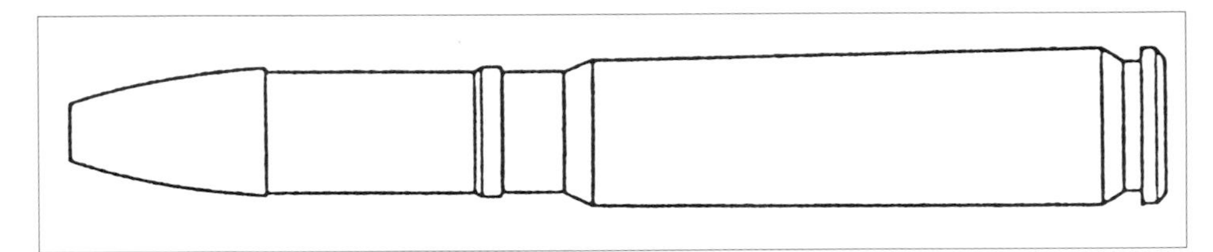

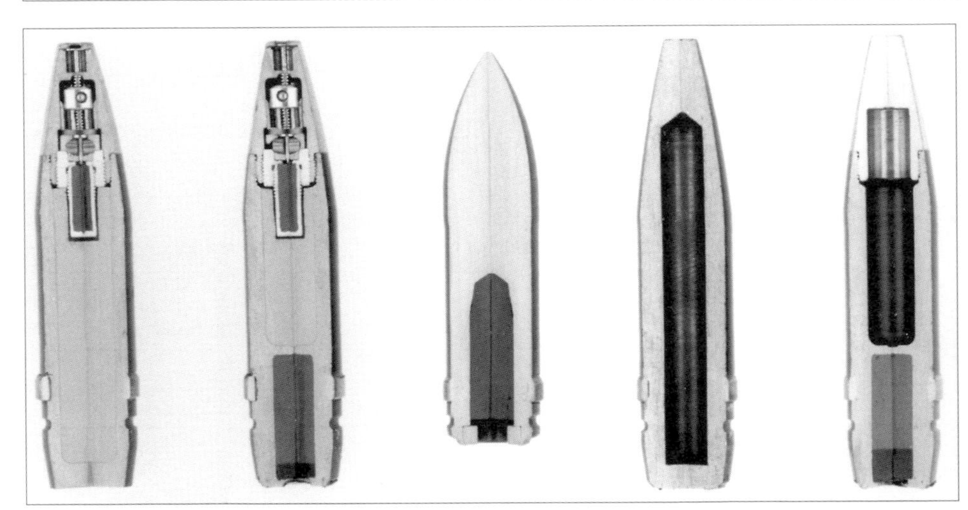

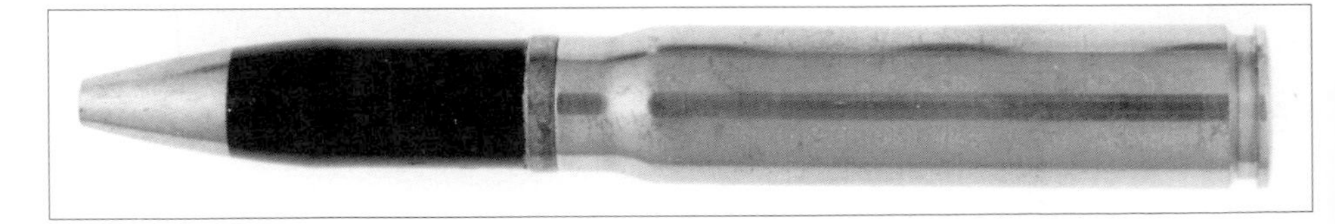

(Right) 20 x 110mm HS404: from the left: HE-I, HE-I-T, AP-T, TP, TP-T.

20 x 110RB

International

Originally known as the 'Oerlikon S' cartridge, this first appeared in the early 1930s and was subsequently adopted all over the world in aircraft cannon. Due to the blowback operation of the original Oerlikon gun, the cartridge case has a rebated rim, the rim being smaller than the base of the cartridge. This allowed the breech bolt to drive the cartridge completely inside the chamber before it fired, rather than having the rim outside the chamber as is more usual. Ammunition is currently manufactured in Belgium, Brazil, France, Greece, Italy, Spain, Serbia and the USA.

DATA

Round length 7.126in (181.0mm)
Case length 4.323in (109.8mm)
Rim diameter 0.874in (22.2mm)
Projectile diameter 0.783in (19.9mm)
Projectile weight 4.30oz (122g)
Muzzle velocity 2,723ft/sec (830m/sec)
Muzzle energy 31,080ft/lbs (42kJ)
Types of projectile available AP, AP-I, AP-T, HE-I, HE-I-T, TP, TP-T

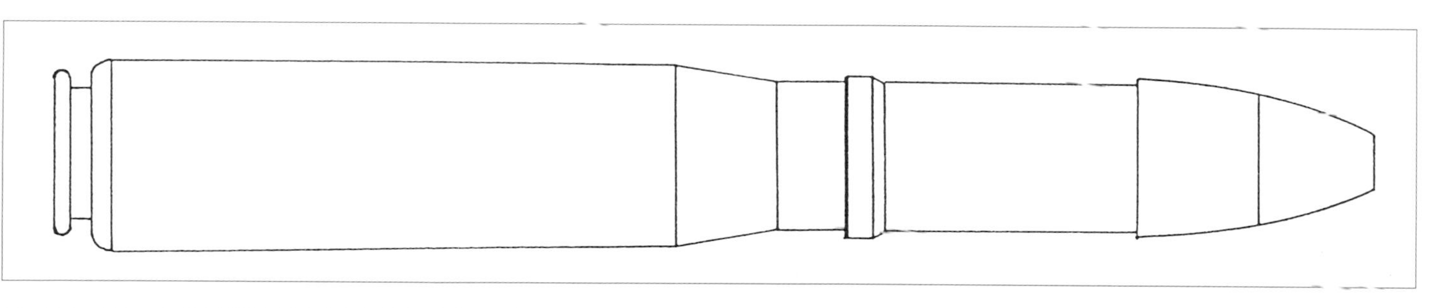

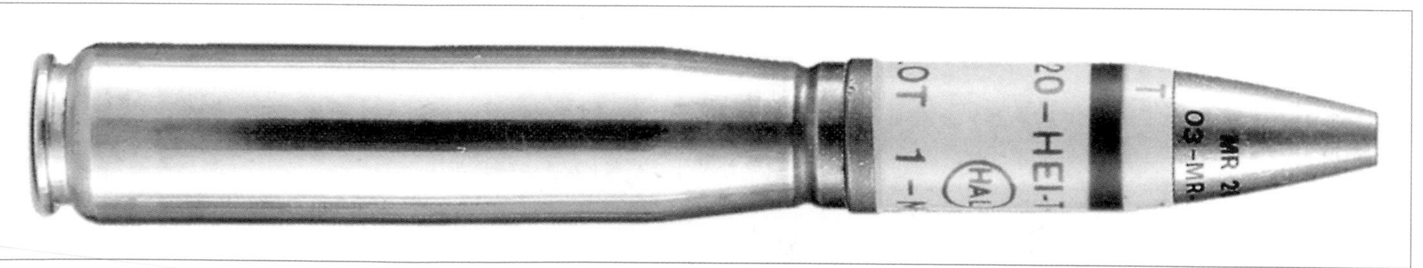

20 x 110 USN

This round was developed in the USA in the 1950s to suit a number of cannon produced for aircraft use. These were then widely sold around the world in US aircraft and ammunition is currently made in Belgium, France, Singapore and the USA. It is also known as the '20mm US Navy Mark 100'.

DATA
Round length 7.283in (185mm)
Case length 4.311in (109.5mm)
Rim diameter 0.807in (20.5mm)
Projectile diameter 0.783in (19.9mm)
Projectile weight 3.88oz (110g)
Muzzle velocity 3,320ft/sec (1,012m/sec)
Muzzle energy 41,662ft/lb (56.3kJ)
Types of projectile available HE-I-T, AP-I-T, TP-T

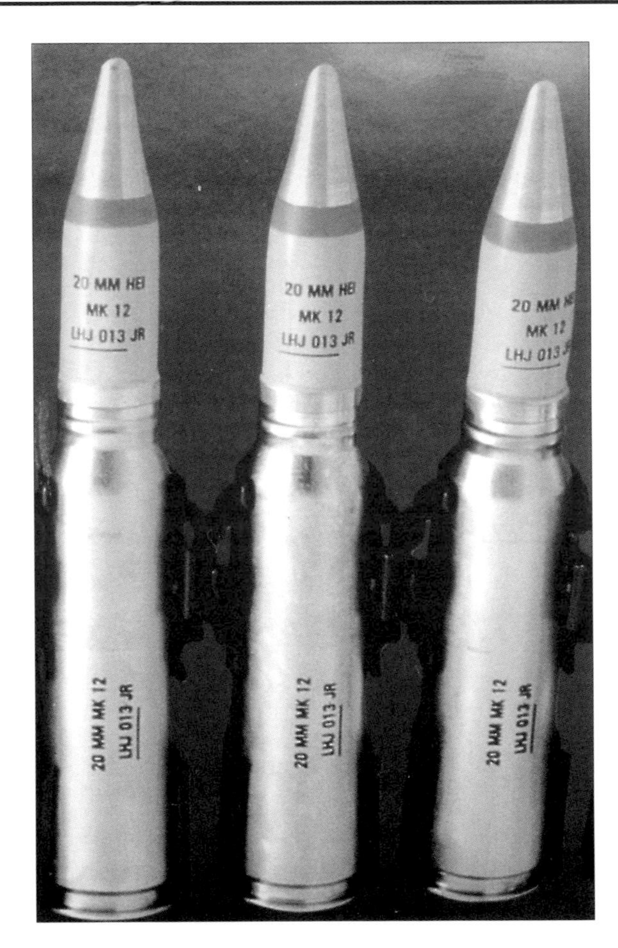

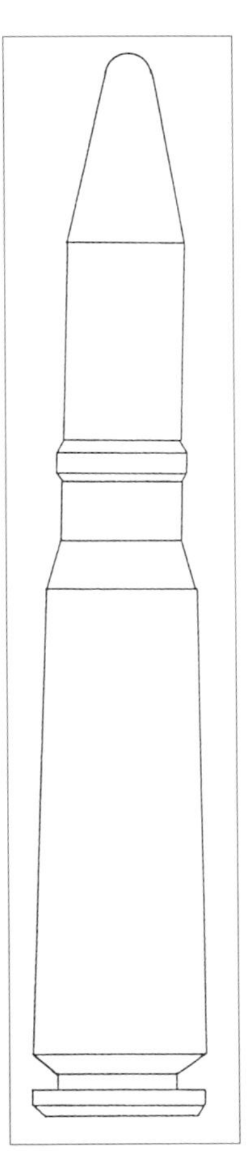

This cartridge was developed in the 1950s by Oerlikon for their KAA cannon and was later adopted for use in other designs. It uses the same projectiles as the 20x139mm HS820 cartridge noted below. Projectile weights vary between 125g and 128g, but the propelling charges are regulated to give a constant muzzle velocity. Ammunition is currently made in France, the Netherlands and Switzerland.

DATA
Round length 7.992in (203.0mm)
Case length 5.067in (128.7mm)
Rim diameter 1.260in (32.0mm)
Projectile diameter 0.783in (19.9mm)
Projectile weight 4.4–4.5oz (125–128g)
Muzzle velocity 3,445ft/sec (1,050m/sec)
Muzzle energy 51,060ft/lb (69kJ)
Types of projectile available HE-I, HE-I-T, AP-I, AP-TTP, TP-T

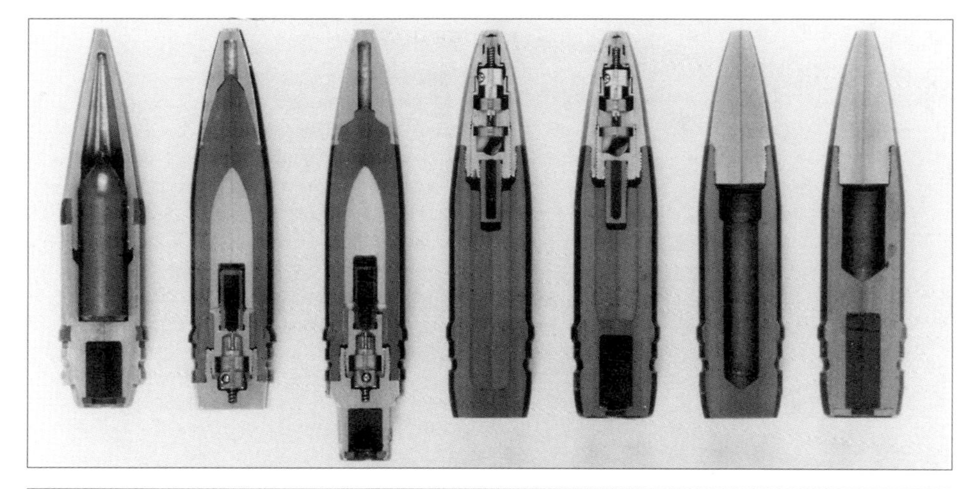

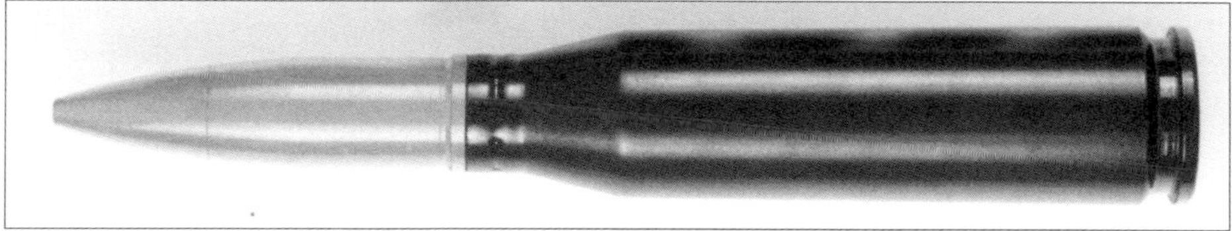

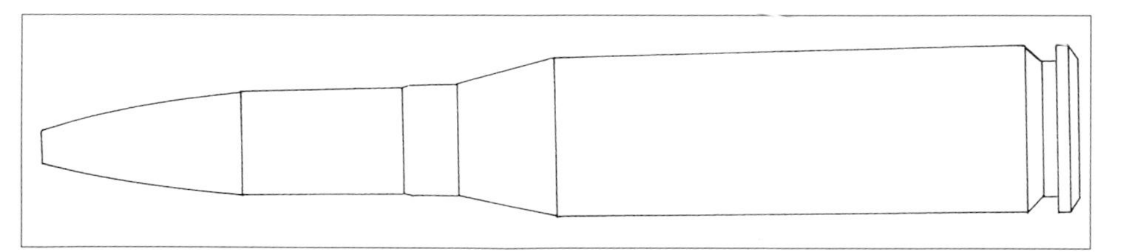

(Top & above) 20 x 128mm KAA: from the left: AP-T, SAP-HE-I, SAP-HE-I-T, HE-I, HE-I-T, TP, TP-T.

This was first produced by Hispano-Suiza just after World War Two as a more powerful replacement for the wartime HS404 round. It was accompanied by a new gun, the HS820, which became one of the most popular 20mm cannon ever made, being widely used as both an aircraft and anti-aircraft weapon. Due to the popularity of the gun and its cartridge, the ammunition was adopted by other designers for their own guns, notably the German Rh202, the French M693 and the American M139. Ammunition is currently made in Argentina, France, Germany, Greece, the Netherlands, Norway, South Africa, Spain and Switzerland.

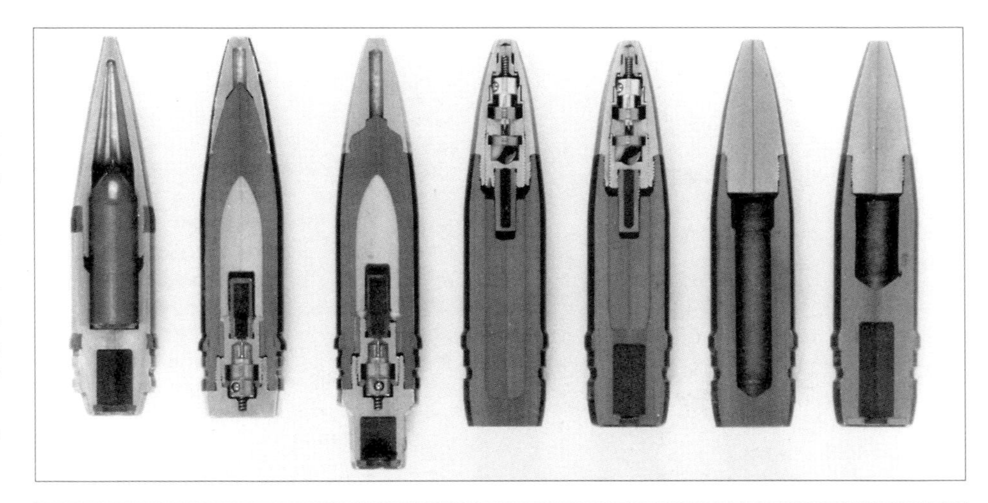

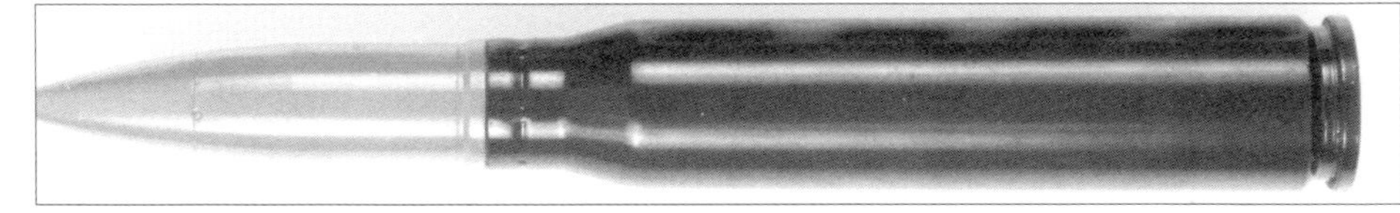

(Top & above) Oerlikon 20 x 139mm rounds: from the left: AP-T, SAP-HE-I, SAP-HE-I-T, HE-I, HE-I-T, TP, TP-T.

DATA
Round length 8.386in (213.0mm)
Case length 5.453in (138.5mm)
Rim diameter 1.118in (28.4mm)
Projectile diameter 0.783in (19.9mm)
Projectile weight ca 4.4oz (125g)
Muzzle velocity 3,412ft/sec (1,040m/sec)
Muzzle energy 53,724ft/lb (72.6kJ)
Types of projectile available AP-T, AP-I-T, HE-I, HE-I-T, MP-T, APDS-T, TP, TP-T

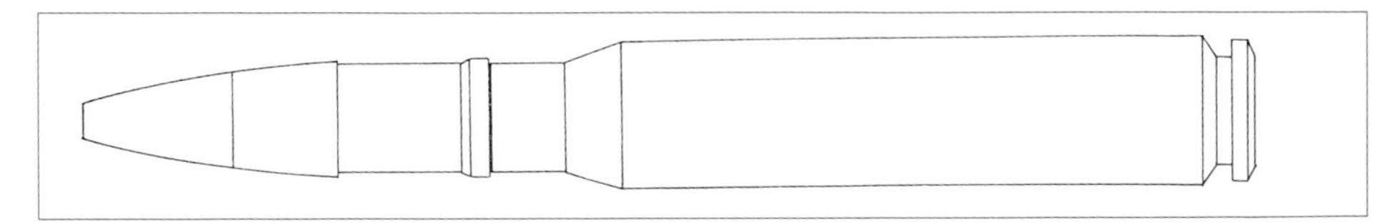

23 x 115 NS

This cartridge was developed during World War Two for the Soviet Nudelmann Suranov NS 23 aircraft cannon. Other cannon designs adopted it in post war years and it was widely distributed throughout the old Warsaw Pact and other Soviet oriented countries. The ammunition for later models of gun was loaded to give a higher velocity than in the original NS 23 gun and marked with a white band on the projectile to indicate this, but it seems that the other countries that adopted the guns and ammunition preferred to use less propellant and accept a slightly lower velocity. Ammunition is currently manufactured by China, Czech Republic, Egypt, Finland, Pakistan, Russia and Serbia.

DATA
Round length 3.882in (198.6mm)
Case length 4.520in (114.8mm)
Rim diameter 1.062in (26.97mm)
Projectile diameter 0.902in (22.93mm)
Projectile weight 6.17oz (175g)
Muzzle velocity 2,428ft/sec (740m/sec)
Muzzle energy 35,446ft/lb (47.9kJ)
Types of projectile available HE-T, HE-I-T, AP-I, AP-I-T, TP, TP-T

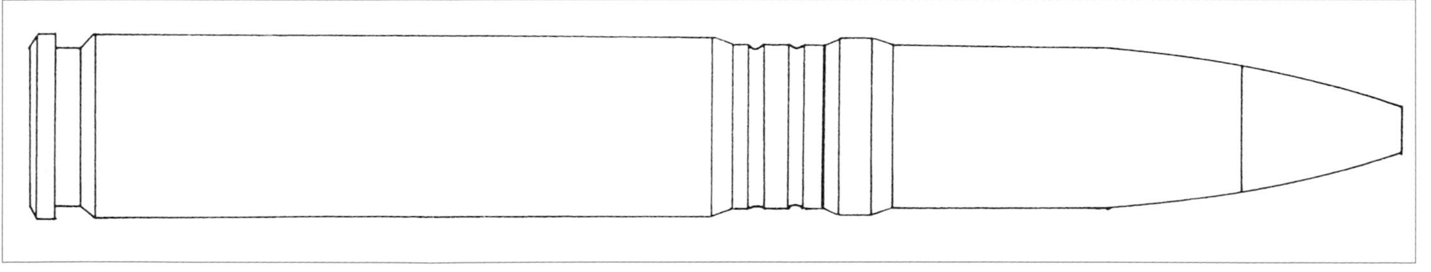

23 x 152B ZU-23

Another wartime Soviet development, notably for use in the 'Sturmovik' ground attack aircraft and its VYa cannon. The cannon was replaced by other designs, but the cartridge was adopted for the ZU-23 cannon, first used as an aircraft gun but then entirely relegated to the anti-aircraft role. In this guise it was widely distributed around the world and ammunition is currently made in Egypt, Iran, Russia, South Africa and Serbia. However, the rounds for the original VYa gun and the later ZU 23 gun are different and are not interchangeable. VYa rounds used a brass case with a screwed-in primer, ZU-23 guns use steel cases and a pressed in primer which cannot be removed.

DATA (ZU-23)
Round length 9.252in (235.0mm)
Case length 5.948in (151.1mm)
Rim diameter 1.305in (33.15mm)
Projectile diameter 0.903in (22.93mm)
Projectile weight 6.52oz (185g)
Muzzle velocity
3,280ft/sec (1,000m/sec)
Muzzle energy
68,450ft/lb (92.5kJ)
Types of projectile available
HE-I-T, AP-I-T

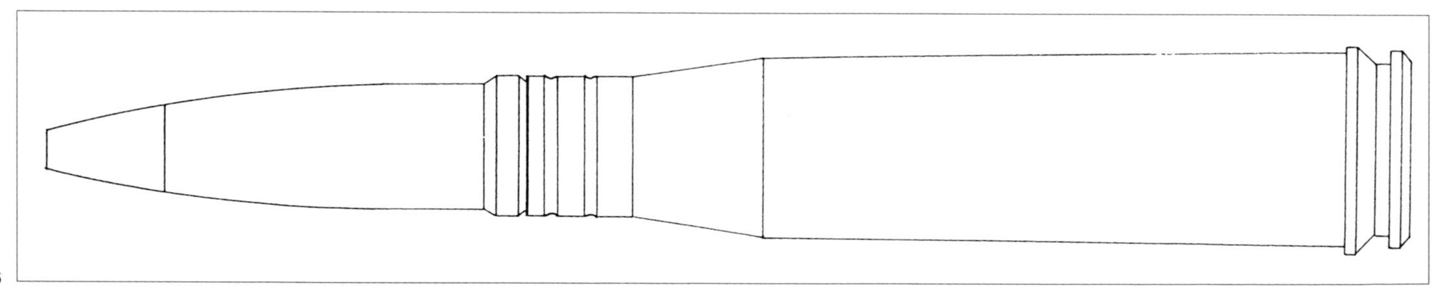

25 x 137 KBA

In the early 1960s the American company Thompson Ramo Wooldridge developed the TRW-6425 25mm cannon as part of the US Army's 'Bushmaster' fighting vehicle programme. In the end the Bushmaster project was shelved and the rights to the gun, which had been designed by Eugene Stoner, were purchased by Oerlikon of Switzerland. The company then made a few modifications and developed this round of ammunition, the result being the KBA cannon. Once the utility of this calibre was recognized, the Bushmaster gun was revived and various other weapons have been developed to fire it.

The necked, steel cartridge case is easily recognisable by the groove below the shoulder, into which the belt link fits. When the cartridge is fired, this is expanded out to make a flush surface, but the mark remains. Projectiles were originally developed by Oerlikon, but other firms, notably Raufoss of Norway and Alliant in the USA, have produced their own designs. Manufacture currently takes place in Belgium, France, Germany, Norway, Switzerland and the USA.

DATA
Round length 8.780in (223mm)
Case length 5.394in (137.0mm)
Rim diameter 1.496in (38.0mm)
Projectile diameter 0.984in (24.99mm)
Projectile weight 6.35oz (180g)
Muzzle velocity 3,610ft/sec (1,100m/sec)
Muzzle energy 80,660ft/lbs (109kJ)
Types of projectile available AP-HC-T, APDS-T, APFSDS-T, FAPDS-T, HE-T, HE-I-T, MP-T, TP, TP-T

(Below) Oerlikon 25 x 137mm rounds: from the left: APDS-T, SAP-HE-I-T, HE-I-T, TP-T.

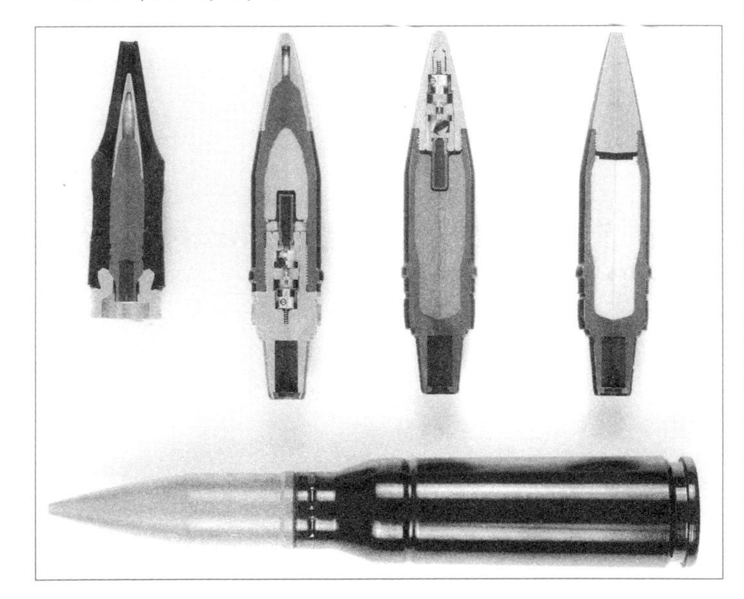

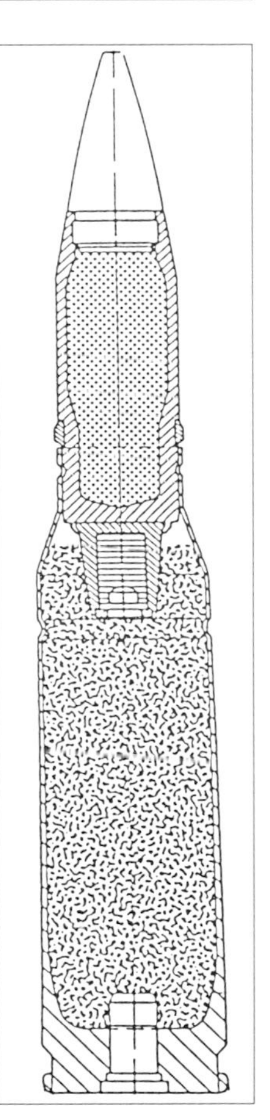

25 x 184 KBB

Switzerland

The Oerlikon KBA cannon was good, but Oerlikon was of the opinion that it could stand a little more power and developed the KBB gun, for which this cartridge was designed. The gun is more or less the same, the improved performance coming from the larger shell and cartridge in conjunction with a longer barrel. So far, no other manufacturer has developed a gun around this cartridge and the only ammunition is made by Oerlikon, but it has considerable potential for the future.

DATA
Round length 11.340in (288.0mm)
Case length 7.244in (184.0mm)
Rim diameter 1.520in (38.60mm)
Projectile diameter 0.980in (24.90mm)
Projectile weight 8.11oz (230g)
Muzzle velocity 3,805ft/sec (1,160m/sec)
Muzzle energy 114,700 ft/lb (155kJ)
Types of projectile available HE-I, APDS-T, AMDS, FAPDS, TP, TP-T, TPDS-T

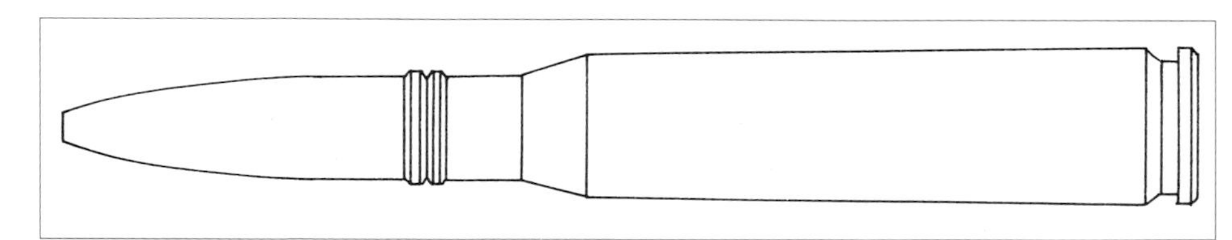

(Above right) 25 x 184mm ammunition for the Oerlikon KBB cannon: from the left: HE-I, TP-T, APDS-T, APP-T.

27 x 145 BK

'BK' stands for 'Bord Kanone' and this round was developed in 1976 for the 27mm Mauser cannon in the Tornado Multi-Role Combat Aircraft. It was later adopted by a number of naval 'close-in weapon' systems for defence against sea skimming missiles. Ammunition is currently manufactured in Italy, Norway and Britain.

DATA
Round length 9.566in (243.0mm)
Case length 5.708in (145.0mm)
Rim diameter 1.358in (34.50mm)
Projectile diameter 1.059in (26.90mm)
Projectile weight 9.17oz (260g)
Muzzle velocity
3,363ft/sec (1,025m/sec)
Muzzle energy
100,862ft/lbs (136.3kJ)
Types of projectile available HE, AP-I, APFSDS-T, MP-T, TP, TP-T

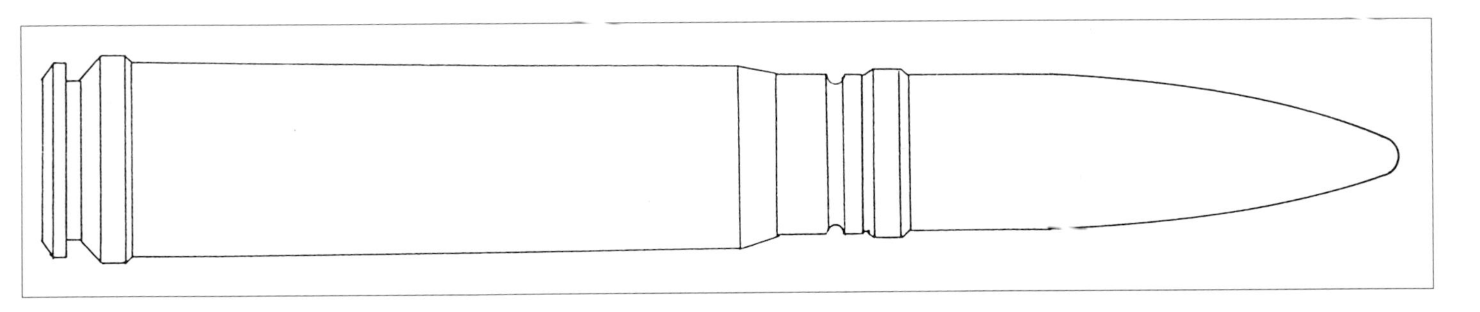

30 x 150B Giat

This is a relatively new round of ammunition which is under development by Giat of France for the M791 cannon fitted to the Rafale fighter. As is obvious from its appearance, it has been evolved from the 30mm ADEN/DEFA round by lengthening the case and increasing the propellant. At the time of writing, only one projectile has been announced, a high explosive shell with base fuze.

DATA
Round length 10.00in (254.0mm)
Case length 5.905in (150.0mm)
Rim diameter 1.358in (34.50mm)
Projectile diameter 1.173in (29.80mm)
Projectile weight 9.7oz (275.0g)
Muzzle velocity 3,363ft/sec (1,025m/sec)
Muzzle energy 106,708ft/lbs (144.2kJ)
Types of projectile available HE-I-T

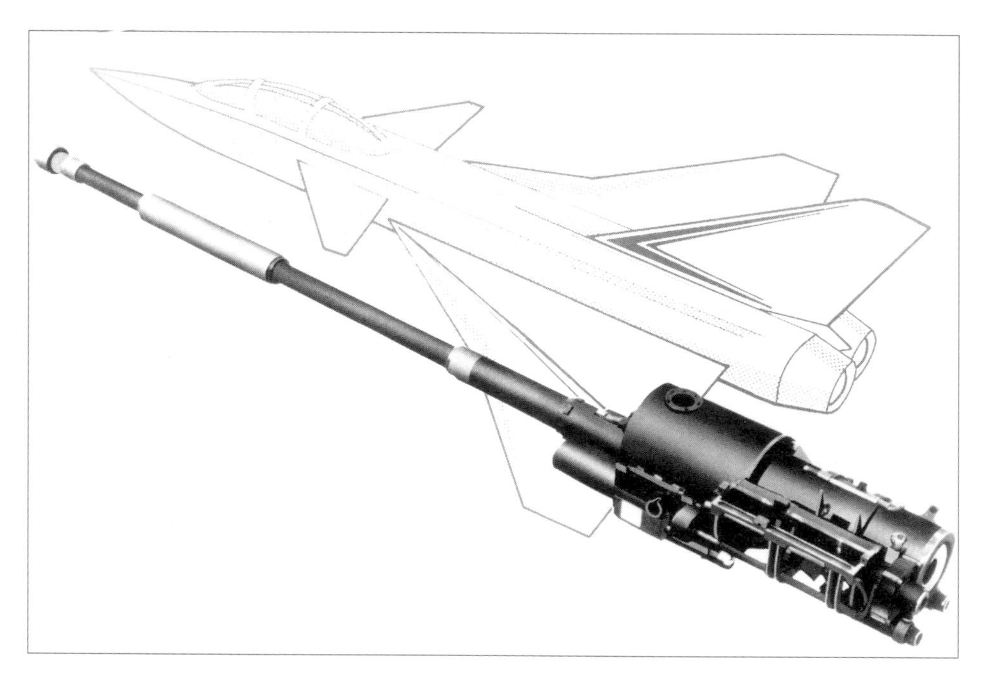

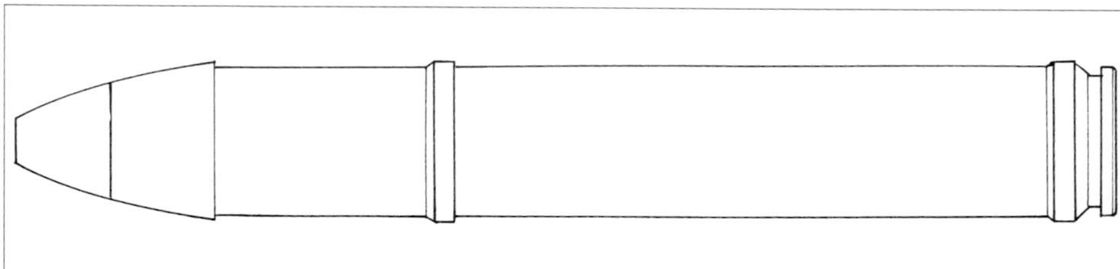

(Top) Giat 30mm (791B) cannon

30 x 155B NR-30

This round came with the Nudelmann-Richter cannon which was fitted to virtually all Soviet fighter aircraft from the mid-1950s. It has been replaced in Russian service by newer cannon designs, but is still in use in China, Egypt and Pakistan, all of whom manufacture ammunition. The case has a substantial belt forward of the extraction groove and in keeping with most Soviet designs, the case is crimped into two grooves around the rear of the projectile behind the driving band.

DATA
Round length 10.433in (265.0mm)
Case length 6.102in (155.0mm)
Rim diameter 1.574in (40.0mm)
Projectile diameter 1.173in (29.80mm)
Projectile weight 14.5oz (410g)
Muzzle velocity 2,559ft/sec (780m/sec)
Muzzle energy 92,278ft/lbs (124.7kJ)
Types of projectile available HE-I, AP-I, TP, TP-T

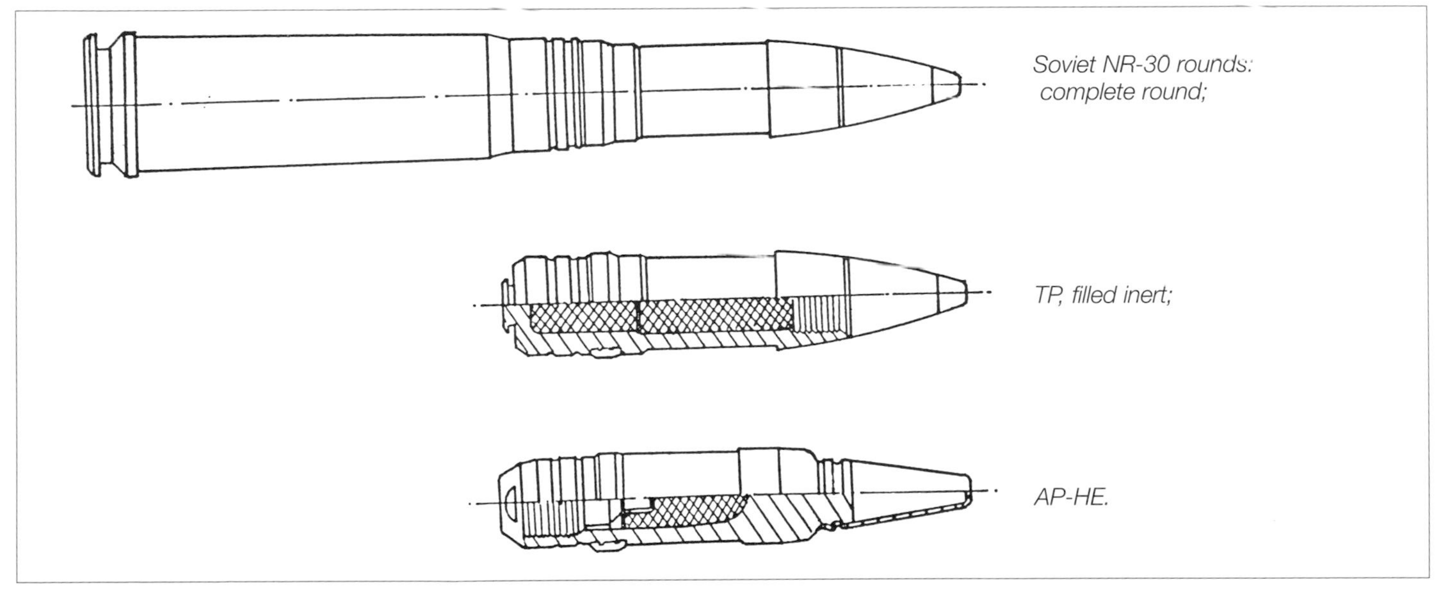

Soviet NR-30 rounds: complete round;

TP, filled inert;

AP-HE.

30 x 113B Aden/DEFA International

The 30mm cartridges for the British ADEN gun and the French DEFA cannon have a long and involved history. Both guns were developed from the Mauser MK213 revolver cannon of 1945, both use short (86mm and 97mm) belted cases and both had poor performance. Subsequently Britain and France collaborated on entirely new designs of gun and ammunition, resulting in the 30 x 113mm cartridge which produced somewhat better performance. Strangely, collaboration extended only as far as case dimensions: French ammunition was loaded to give velocities between 760 and 820m/sec, while British rounds achieved figures varying from 600 to 800 m/sec.

Current designs for the two guns still differ, and although the velocities are now closer, most British ammunition being 785m/sec and French varying between 775 for APHE and 810 for HE-I. Both countries use a similar belted case, the French being lacquered steel and the British brass. The HE projectiles of both countries are notable for being hemispherical based and drawn from high tensile steel to give extremely high capacity for the explosive. The British HE-I shell, for example, is loaded with 48g of explosive, some 50% more than a conventional cast steel shell could hold.

Equivalent rounds are also manufactured in Argentina, Belgium, Brazil, Egypt, Finland, Greece, India, Norway, Singapore, South Africa and the USA.

DATA
Round length 7.874in (200.0mm)
Case length 4.449in (113.0mm)
Rim diameter 1.311in (33.30mm)
Projectile diameter 1.181in (30.00mm)
Projectile weight 8.64oz (245g)
Muzzle velocity 2,657ft/sec (810m/sec)
Muzzle energy 59,570ft/lbs (80.5kJ)
Types of projectile available HE-I, AP-I-T, MP-T

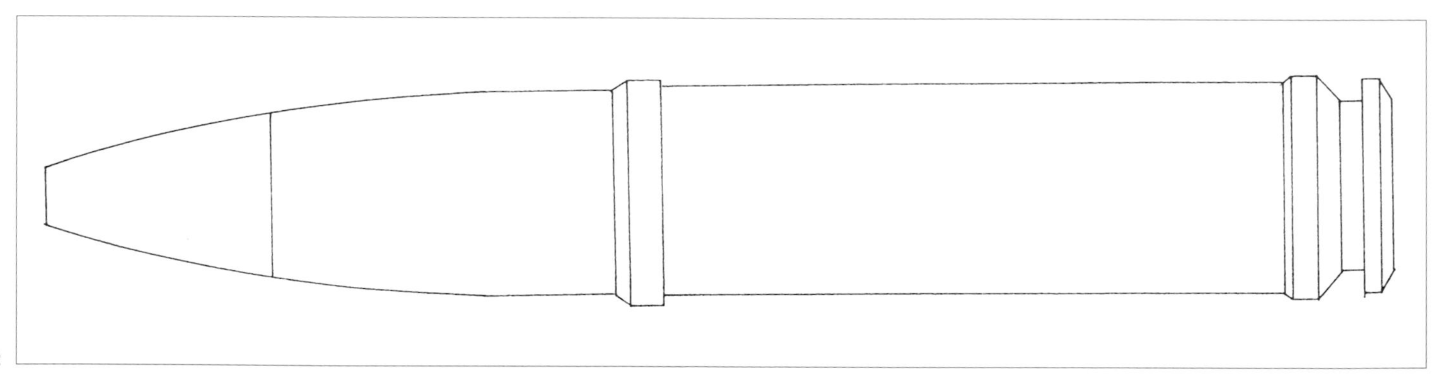

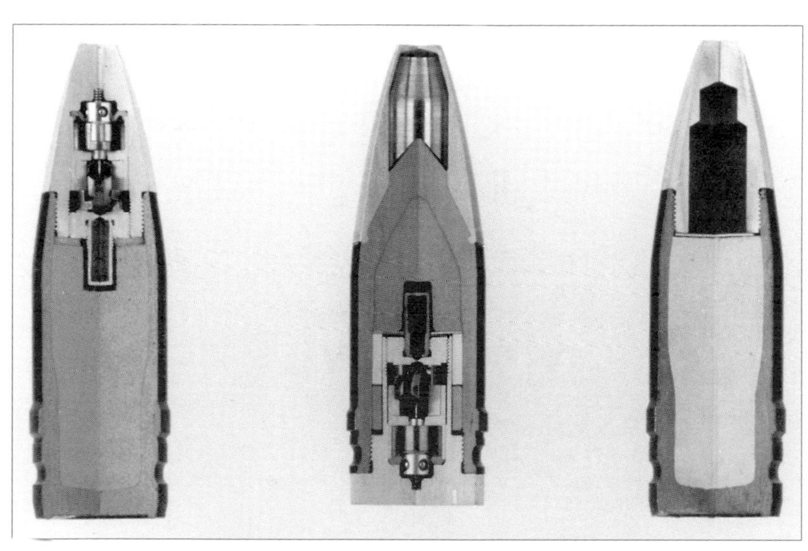

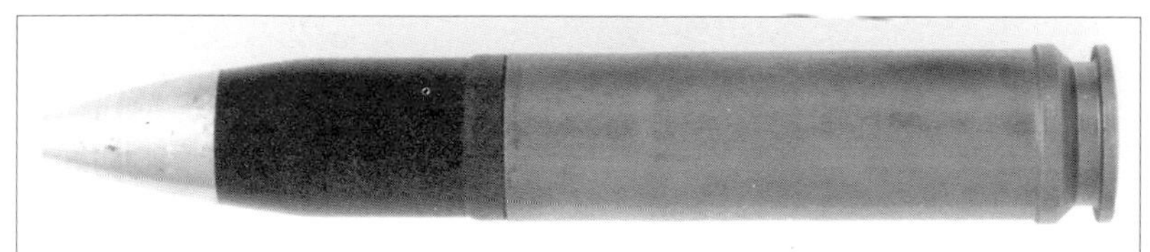

(Top) Oerlikon 30 x 113B rounds: from the left, HE-I, SAP-HE-I, TP.

(Right) 30mm DEFA shells; left HE-I, right TP.

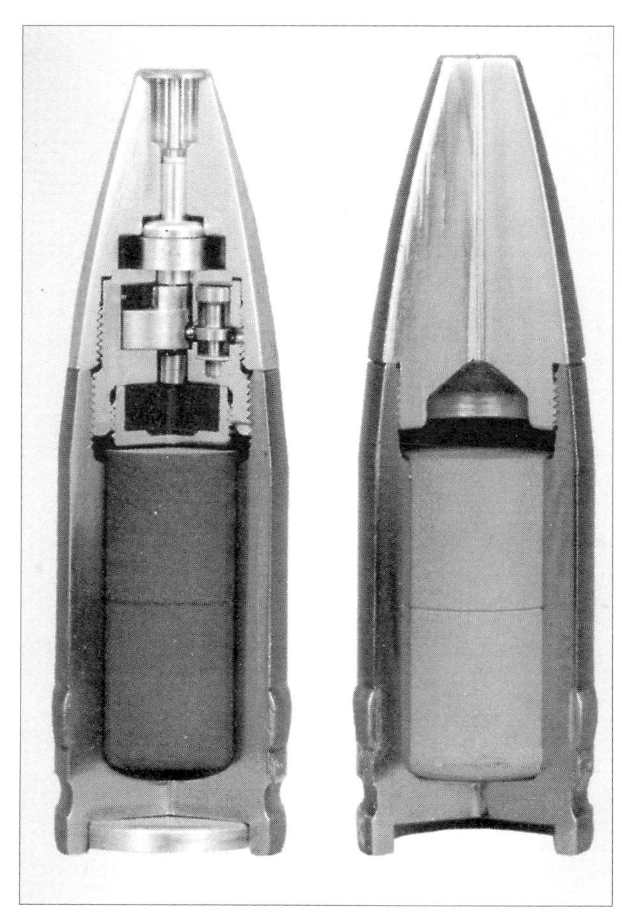

30 x 170 Oerlikon KCB

This powerful round was first developed by Hispano-Suiza for their HS831L cannon in the 1950s and when Hispano was absorbed by Oerlikon the cannon went through some slight changes to appear as the KCB. The cartridge was also taken as the starting point for the British Rarden cannon, but this development resulted in so many changes that it is barely recognisable as the same cartridge, although KCB rounds can be fired from Rarden guns. The KCB ammunition is used in a number of weapons designed for light anti-aircraft or armoured vehicle roles in several countries. Ammunition is currently made in Argentina, Britain, Finland, France, Greece, the Netherlands, Norway, South Korea and Switzerland.

(Above) Oerlikon 30 x 170mm rounds: from the left, HE-I, HE-I-T, SAP-HE-I, TP, TP-T.

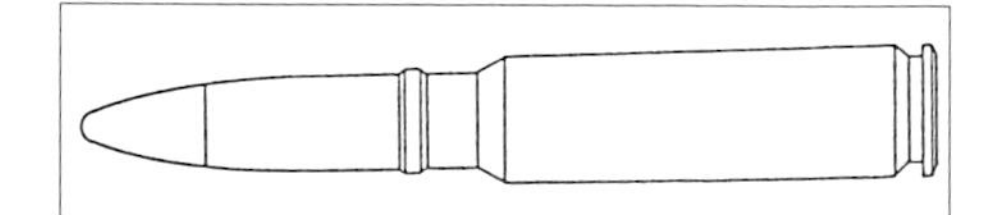

DATA

Round length 11.220in (285mm)
Case length 6.693in (170mm)
Rim diameter 1.173in (29.80mm)
Projectile diameter 1.173in (29.80mm)
Projectile weight 12.7oz (360g)
Muzzle velocity 3,543ft/sec (1,080m/sec)
Muzzle energy 195,212ft/lbs (263.8kJ)
Types of projectile available HE-I, HE-I-T, TP, TP-T, MP-T

30 x 173 Oerlikon KCA

The 30mm Oerlikon KCA revolver gun, with a rate of fire of 1,250 rounds per minute, is one of the most potent aircraft cannons in existence and was introduced in the early 1980s. The cartridge has since been adopted for other weapons, such as the American GAU 8/A Gatling gun and the Mauser MK30, and will doubtless appear in more in the future. The original rounds were quite conventional, but recent development has produced APDS rounds with a muzzle velocity of 1385m/sec and a 'Frangible Missile Piercing Discarding Sabot' round produced for the Netherlands Navy's 'Goalkeeper' close-in weapon system for use against sea skimming missiles. Ammunition is currently made in the Netherlands, Switzerland and the USA.

DATA
Round length 11.417in (290mm)
Case length 6.811in (173.0mm)
Rim diameter 1.732in (44.0mm)
Projectile diameter 1.177in (29.90mm)
Projectile weight 12.8oz (363g)
Muzzle velocity
3,395ft/sec (1,035m/sec)
Muzzle energy 143,589ft/lbs (194.03kJ)
Types of projectile available AP-I, HE-I, APDS-T, FMPDS-T, TP-T

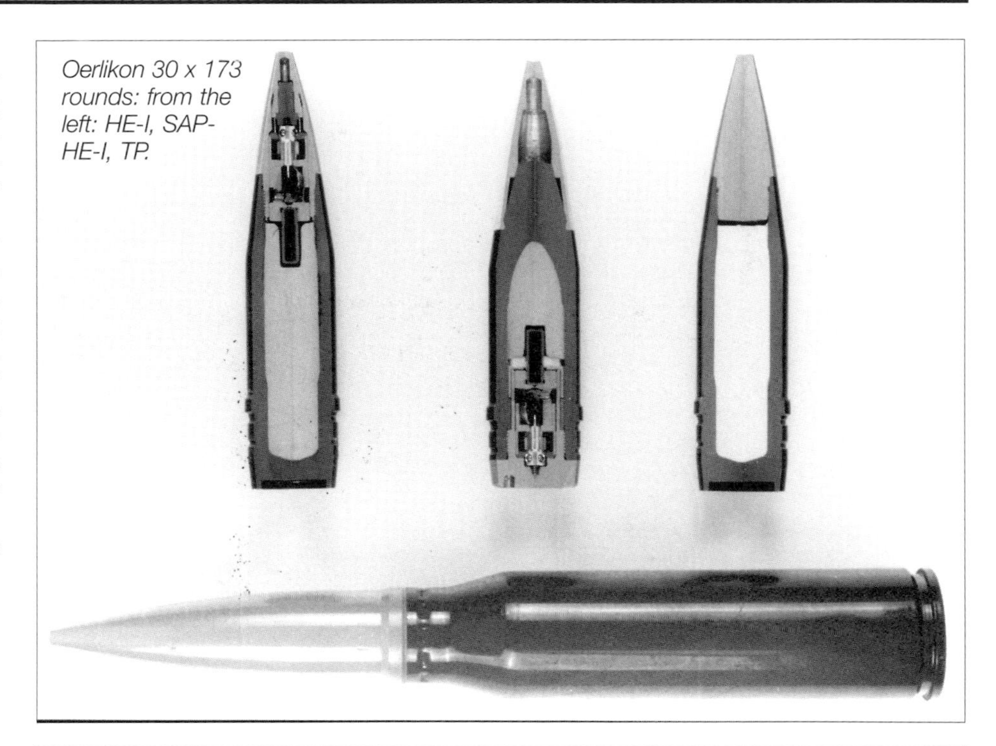

Oerlikon 30 x 173 rounds: from the left: HE-I, SAP-HE-I, TP.

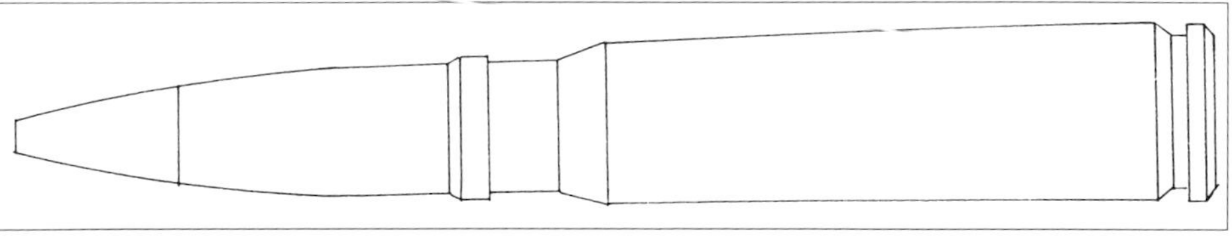

Cased Telescoped Ammunition is still under advanced development and the data given here should be regarded as provisional. If and when the weapon enters service the ammunition may well have changed its specification, though the original principle will have been adhered to.

As described in the text, the CTA round consists of a projectile wholly contained inside the cartridge case, together with the propellant, the whole being closed by a plastic sealing cap around the nose of the projectile, so that the nose is visible in the mouth of the case. The case has to carry a means of ignition – usually an electric primer – and also the small booster charge which launches the projectile from the case before the main charge fires.

There are now two different types of CTA 45mm ammunition undergoing evaluation: the European design uses a composite material case with a slightly convex base to suit the breech mechanism. The American 'COMVAT' (Combat Vehicle Armament Technology) round uses a light metal case with a convex base and with an extraction groove and rim. The projectiles are of more or less conventional form, though the design detail necessary to fit them into the case and conform to the loading technique shows some departures from the usual practice.

DATA
Round length 12.00in (305mm)
Case length 12.00in (305mm)
Rim diameter 2.756in (70mm)

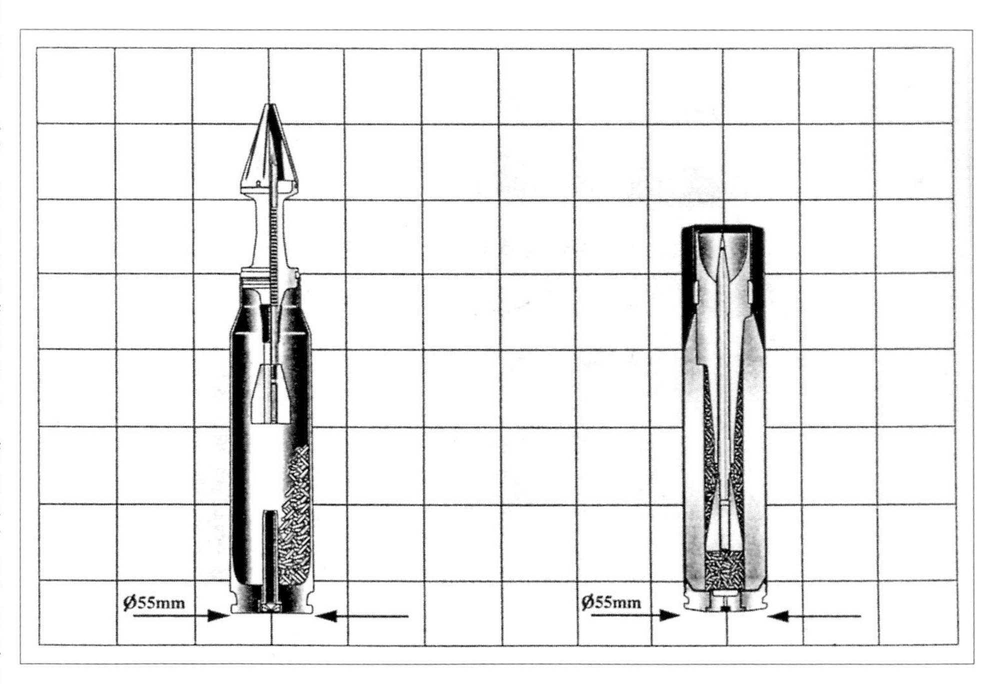

Comparison of a conventional 35mm cartridge (left) and a 45mm CTA cartridge (right) showing that a 45mm CTA is the same diameter and is shorter, but actually has twice the power of the 35mm round.

Projectile diameter 1.771in (45mm)
Projectile weight ca 1.5lbs (700g)
Muzzle velocity >3,608ft/sec (1,100m/sec)

Muzzle energy ca. 307,000ft/lbs (415kJ)
Types of projectile available: HE-I-T, APFSDS, TP-T, TPFSDS, FAPDS-T

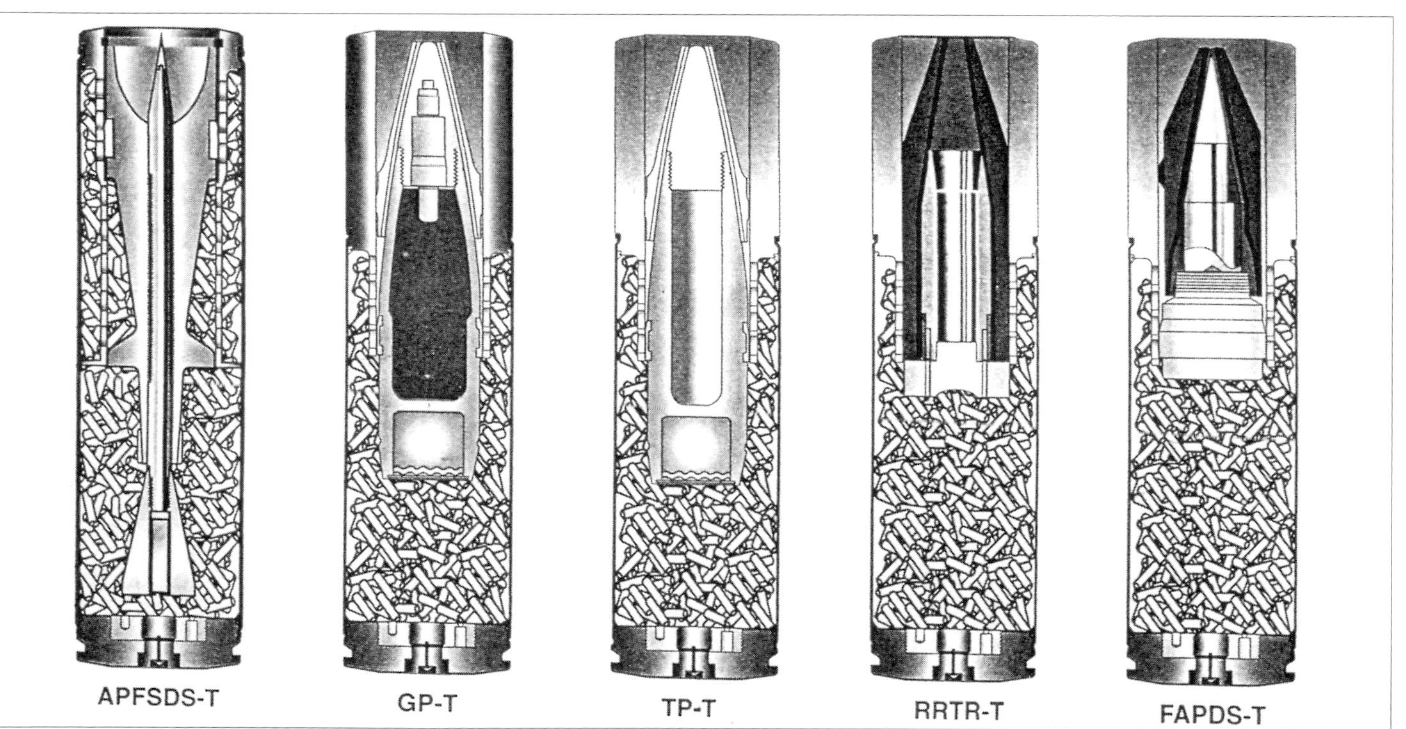

Grenades

Grenades are frequently regarded as a simple, safe and relatively harmless proposition, but the fact is that they are complex and extremely dangerous devices which deserve more understanding and respect than they usually receive.

The specification for a hand grenade would amaze most people in the number of apparently conflicting conditions it lays down. The grenade must be safe to transport, carry and throw, but must then immediately arm itself and become capable of detonating at the end of the thrown distance. Should the thrower be shot as he is winding up to make his throw, then the grenade must be harmless as it falls from his hand. It should unerringly kill or injure anyone within 10 yards of its burst, but not present any hazard to anyone 15 yards from it. It must operate irrespective of the attitude in which it falls or the hardness or softness of the ground upon which it falls and so on. A grenade also has to be cheap, easily manufactured by any workshop, not use scarce or strategic materials and have a shelf life of 10 years. These latter demands are best explained by simply remembering that between July 1915 and November 1918 Britain turned out 68,329,245 Mills grenades, which equates to 56,100 grenades per day of only one pattern.

The Mills, or No 36 grenade, with a base disc attached to permit firing from a cup attachhment on the muzzle of a rifle. The lever and safety pin hold up the central striker against a spring. Beneath it is a cap, with a fuze leading to the detonator on the right side.

The Mills grenade had a long run; it appeared in July 1915 and was still in use in the Korean War and the reason was simply the demands enumerated above. Many and varied were the designs which appeared before and during World War Two, but none of them were fit to replace the Mills' design.

Anti-personnel hand grenades can be broadly divided into two classes, offensive and defensive; the difference lying in the nature of the fragmentation. An offensive grenade is one which is used when attacking and the significant feature is that the fragments must not spread so far as to injure the man who threw it who will, in most cases, be standing up and advancing ready to take advantage of the grenade's effect. The defensive grenade, on the other hand, is to be thrown by a man defending a position and therefore behind some sort of protective cover or in a trench or rifle pit; in this case he throws and ducks and the fragments can go farther so as to provide a large lethal area. This distinction began to be appreciated during World War Two and a variety of grenades appeared in both

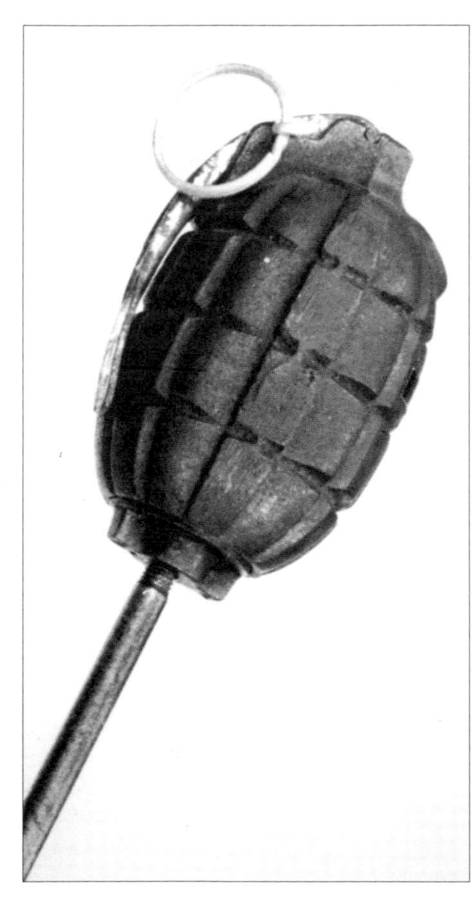

An earlier Mills design, the No 23 grenade, with a rod attached to allow it to be fired from a rifle.

categories. There were also attempts to devise grenades which would fulfil both roles, by having the basic grenade with a thin body, giving small fragments over a small area and then providing a heavy cast iron fragmenting sleeve to slip over the body when required in the defensive role and provide much heavier and wider ranging fragments. Some of these were successful, others not; but they formed the basis of later designs which, as will be seen, are thoroughly effective in both roles.

Hand grenades were very quickly followed by rifle grenades, in an effort to increase the effective range beyond the 20–30 yards which could be reached by the average soldier throwing a grenade. The first models used a steel rod on the base of the grenade which fitted into the rifle barrel and was launched by firing a blank cartridge. These were effective, but soon ruined the rifle for ordinary shooting; the sudden high pressure set up in the barrel as the gas met the rod and began to move the grenade soon bulged the barrel. Next came cups fitted to the muzzle from which ordinary hand grenades could be launched and these remained in use throughout World War Two. Next came barrel extensions which could be clamped on to the rifle's muzzle and over which the hollow tail of a suitable finned grenade could be slipped, again launched by means of a blank cartridge. One or two designs had holes in the middle of the grenade which allowed the use of a ball cartridge, thus removing the need to empty and reload the rifle, but the drawback here was the enormous space demanded for a training ground: the bullet could go for miles at the elevation needed to launch a grenade.

Eventually the design of rifle grenades has settled on a standardised rifle muzzle of 22mm external diameter and a standardised tail unit which fits over it and most modern rifle grenades are built to this NATO standard, whether or not they are actually used by NATO forces. The problem of unloading the ball cartridge to insert a grenade launching blank cartridge and, in gas operated automatic rifles, of remembering to turn the gas regulator so as to shut off the automatic function and thus use all the gas for launching, has been removed by the arrival of the 'bullet trap' and the FN 'Bullet Thru'™ grenades, which permit launching by means of whatever round of ammunition happens to be loaded in the rifle.

Rifle grenades have been under considerable pressure over the past 25 years or so from the 40mm grenade launched from a special 40mm weapon. These generally have greater range, are more accurate and are easier to handle. Against that is the small size of the grenade and its limited area of effectiveness when compared to the 'traditional' rifle grenade, though the arrival of automatic grenade launchers – virtually 40mm machine guns – provides a high rate of fire which compensates for the limited lethal area by simply drowning the target in explosions. There is, to some observers, a further drawback to this weapon, which is its relatively flat trajectory which makes it very difficult to deal with a target concealed behind any sort of frontal cover or in a trench. Here the 50–60mm mortar has a decided advantage and a great deal more effect at the target.

The grenade, whether hand or rifle, began its career as a simple anti-personnel weapon, but it soon became obvious that there could be other applications. Given the conditions of World War One, it is hardly surprising that gas grenades soon appeared, followed by smoke grenades. Rifle grenades firing coloured stars for signalling and white flares for illuminating the front were next and towards the end of the war one or two anti-tank grenades appeared, though they saw little use.

During the 20 years of peace between the two World Wars, almost all the wartime grenades vanished; only the British Mills and the German stick grenade survived, both in improved forms. New smoke and signal grenades were developed, but the gas grenade was abandoned and nothing was achieved in the field of anti-tank grenades. World War Two brought new designs, some of which lingered after the war, but in the 1960s new patterns began to appear and since that time there has been a constant flow of new models, some of which have survived and some not. The following entries present a selection of grenades that are currently in use.

RGN (sectioned drawing)

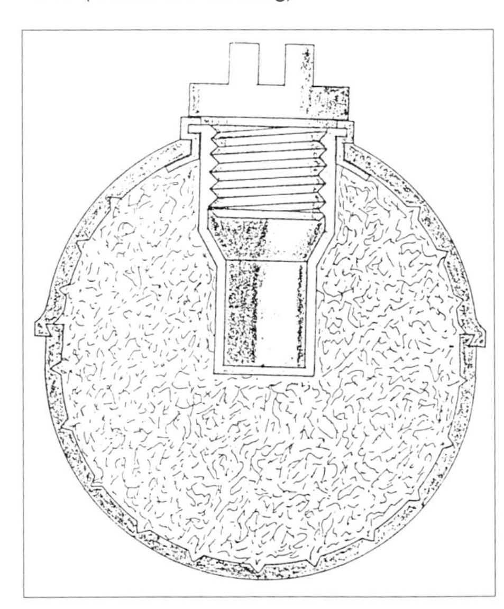

As is usually the case with Soviet designs, nothing is known about the background of this grenade, although there have been suggestions that it originated in Bulgaria and was then adopted by the USSR in the 1970s.

The Russian description calls this an offensive/defensive grenade, but the construction leans more towards the offensive role. The spherical casing is of aluminium alloy and, as can be seen from the sectioned drawing (left), is serrated on the inside surface so as to control the break up into fragments. The usual heavy serration on the outside of hand grenades does nothing to assist fragmentation; it is intended to facilitate gripping the grenade with a muddy hand. But serration on the internal surface offers a degree of control over the size of the fragments and also ensures that the entire surface breaks up into similarly sized pieces. However, since the material of this grenade is relatively light, the fragments will have a very high initial velocity, but will very quickly lose their momentum, so that the lethal area is kept within a radius of 10 metres.

The fuzing arrangement is unusual in combining both time and impact operation. The usual sort of lever and mousetrap arm mechanism is used and when thrown the striker fires an igniter which lights three pyrotechnic delay units. Within the first one and a half seconds of flight, two of these delays burn away and unlock an impact plunger. If the grenade strikes the ground, or a target, within three to four seconds of flight, then this plunger will move and fire a detonator to initiate the main firing of explosive. If, however, the grenade is still in flight at the end of the three to four second delay, then the third pyrotechnic unit will ignite a separate detonator and set off the explosive charge. Thus the grenade can be thrown directly at a target and function on impact, or it can be thrown high into the air to descend and burst over the target, relying upon the skill of the thrower to achieve the desired effect.

DATA
Weight 10.2oz (290g)
Diameter 2.40in (61mm)
Filling 3.4oz (97g) RDX/Wax
Lethal radius 8–10 yards/metres
Delay time Impact 1.8s; Time functioning 3.2s

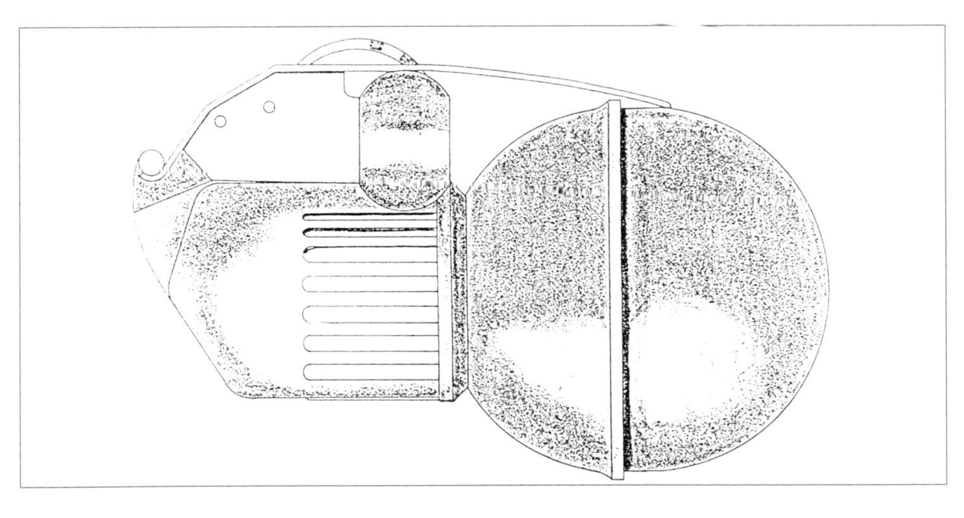

M26

USA

The US Army went through World War Two using a grenade which it had adopted from the French in 1917 and scarcely modified since. It had its defects, principally faulty and erratic fragmentation and by 1945 work was in progress on an entirely new design which appeared after the war as the M26. It was thereafter adopted by several countries, among them Great Britain, where it was labelled Grenade, Hand, L2A1.

The M26 grenade was designed so as to produce the maximum number of optimum sized fragments and distribute them efficiently in all directions. The body is made of thin sheet steel in two hemispheres. Inside this body is a shaped coil of notched, spring steel, square section wire. This is placed into the lower hemisphere and the top hemisphere fitted on top and crimped into position. The interior is then filled with TNT, leaving a central channel into which the igniter set will fit. The hole at the top of the body is threaded.

The igniter set is commonly known as the 'mouse trap' type. There is a diecast metal holder which is shaped to screw into the top of the grenade and with a hole in it through which a delay unit, detonator and cap are fitted. There is a hinged flap, with a raised firing pin, attached to one side of the holder and fitted with a strong spring which forces it to lie on top of the holder. A pressed steel handle has a lip on one end which locks beneath the edge of the holder.

The igniter set is prepared by first pulling back the striker flap against its spring, hooking the lever on to the lip and pressing it down so as to keep the striker flap flat. The lever is then locked in place by passing a split pin through two lugs on the holder. The igniter, delay and detonator are then fitted in place. This is achieved during initial assembly and the lever and holder supplied with, but separate from, the filled grenade body. When the soldier intends to use the grenade he screws the igniter set into the top of the grenade and it is then ready for use.

To use the grenade, it is grasped so as to hold the lever against the grenade body. The safety pin is pulled out and the grenade thrown. As it leaves the hand, so the lever is released and the pressure of the spring on the striker flap causes it to be thrown off, thus allowing the flap to swing across and drive the firing pin into the cap of the detonator assembly. The cap ignites a short length of delay fuze and after four or five seconds this fuze fires the detonator, which initiates the TNT inside the grenade.

It is also possible to use the M26 as a rifle grenade; an adapter, in the shape of a tube with fins at the rear end and claws at the front, is fitted by simply snapping the grenade into the claws. An inertia catch on one of the claws is slipped over the grenade handle and the tail unit is fitted over the muzzle of the rifle. A blank cartridge is loaded and finally the safety pin is withdrawn from the grenade so that the lever is held only by the inertia clip. When the grenade is launched, this clip flies back and releases the lever, after which the action is exactly the same as before. The difference is that the grenade is moving faster and therefore can cover more distance during the burning time of the delay.

DATA
Length 3.89in (99mm)
Weight 1lb (454g)
Diameter 2.244in (57mm)
Fuze delay 3–7 seconds
Lethal radius 10 yards/metres

M 26 A2
H.E.
FRAGMENTATION
COMP B.
LOT:IMI 0-75
DELAY 4,5 SEC

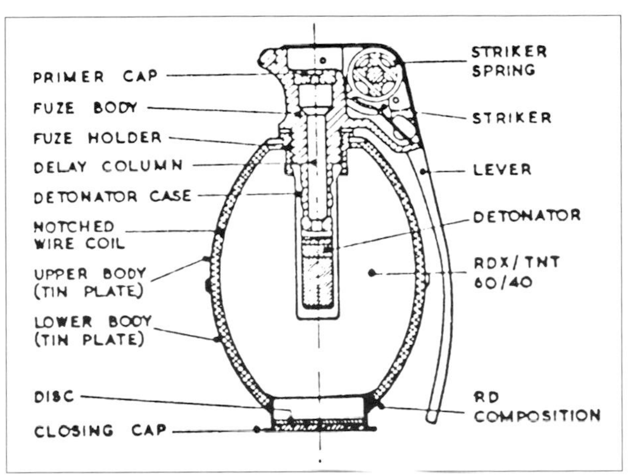
PRIMER CAP
FUZE BODY
FUZE HOLDER
DELAY COLUMN
DETONATOR CASE
NOTCHED WIRE COIL
UPPER BODY (TIN PLATE)
LOWER BODY (TIN PLATE)
DISC
CLOSING CAP
STRIKER SPRING
STRIKER
LEVER
DETONATOR
RDX/TNT 60/40
RD COMPOSITION

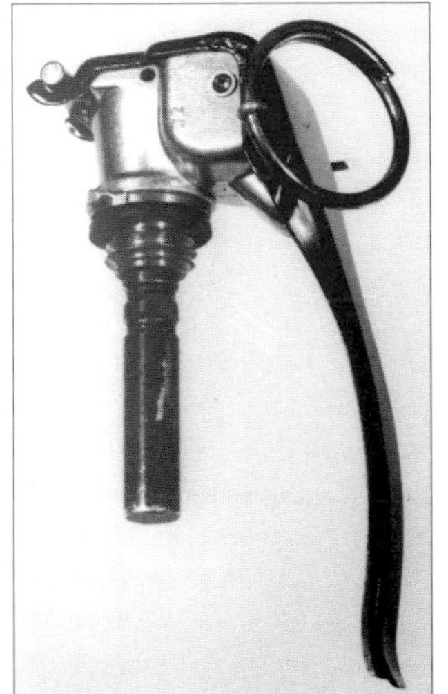

Diehl DM51

This has been the standard German hand grenade for several years and is of the type that can be modified by the user to act either as a defensive or offensive grenade according to the immediate requirement. It consists of two independent units: the hand grenade itself and a fragmentation sleeve.

The grenade body is hexagonal in shape and made from a plastic material. The top and bottom are shaped into a form of bayonet joint and the top is also threaded to take the normal mousetrap fuze assembly. The grenade is filled with PETN high explosive.

The fragmentation sleeve is also of plastic, round and ribbed on the outer surface so as to afford a good grip for the thrower, but hexagonal inside so as to fit closely over the hand grenade body. The sleeve is hollow and the space is filled with 6,500 steel balls, each about 2mm in diameter. At the bottom of the sleeve is a rotatable baseplate.

The hand grenade may be used as it stands, as an offensive or blast grenade, since it produces a great deal of concussion but very little fragmentation. To turn it into a defensive grenade the fragmentation sleeve is slipped over the hand grenade body and locked in place by a half turn of the baseplate. The grenade can then be thrown and, on detonation, provides a shower of high velocity fragments as the steel balls are blown in all directions.

The hand grenade body can also be connected to another grenade body by means of the bayonet catches and a column of grenades can be easily built up to make a demolition charge, firing it by inserting the normal fuze into the topmost grenade.

DATA

Length, fuzed 4.2in (107mm)
Weight, fuzed 15.3oz (435g)
Diameter 2.244in (57mm)
Weight of fragments 9.17oz (260g)
Weight of explosive 2.1oz (60g)

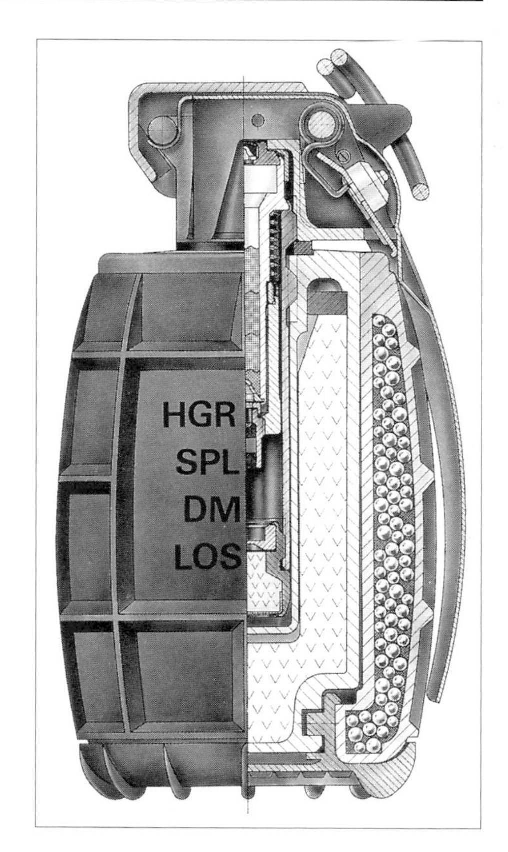

HGR/DN

40mm Arges HE 92

This is a high explosive fragmentation grenade designed to be fired from a 40mm launcher and is one of many which have been developed from the original American idea of the 1960s.

The 40mm launcher relies upon a little used ballistic system to provide sufficient energy to send a grenade for several hundred metres without breaking the shoulder of the man firing the weapon. The idea was developed in Germany in 1944 and is called the `high-low pressure' system and the heart of the matter lies in the cartridge.

The cartridge case is divided into two compartments, a small one surrounding the percussion cap and filled with smokeless powder and the larger one is simply the rest of the cartridge case, closed by the base of the grenade. The smaller compartment has a number of carefully proportioned holes in it. When the cap is fired, the smokeless powder explodes and generates high pressure gas inside the small compartment. This is allowed to pass through the regulating holes and into the larger compartment where it expands to produce a propulsive effort but at a much lower pressure. The advantage is that the propellant burns more efficiently and regularly at high pressure, but the weapon becomes more controllable if the actual pressure driving the grenade is kept low but is able to produce a more sustained push.

The grenade is a cylinder with a blunt nose and a fuze with self destruction element in the rear end. The body is double skinned, with about 1,000 2mm steel balls packed into the space between the inner and outer skins. Inside the inner skin is the filling of high explosive. The fuze is armed on firing and will detonate the grenade if it strikes anything, but should the impact element fail, as, for example, if it lands in soft mud or sand, then after 15 seconds the self destruction element will detonate it.

DATA
Calibre 40mm (1.57in)
Length, complete 4.0in (102mm)
Weight, complete 9.3oz (265g)
Weight, grenade 6.7oz (190g)
Grenade filling 0.98oz (28g) RDX
Fragments ca 1,000 2mm balls
Muzzle velocity 250ft/sec (76m/sec)
Maximum range 437 yards (400m)

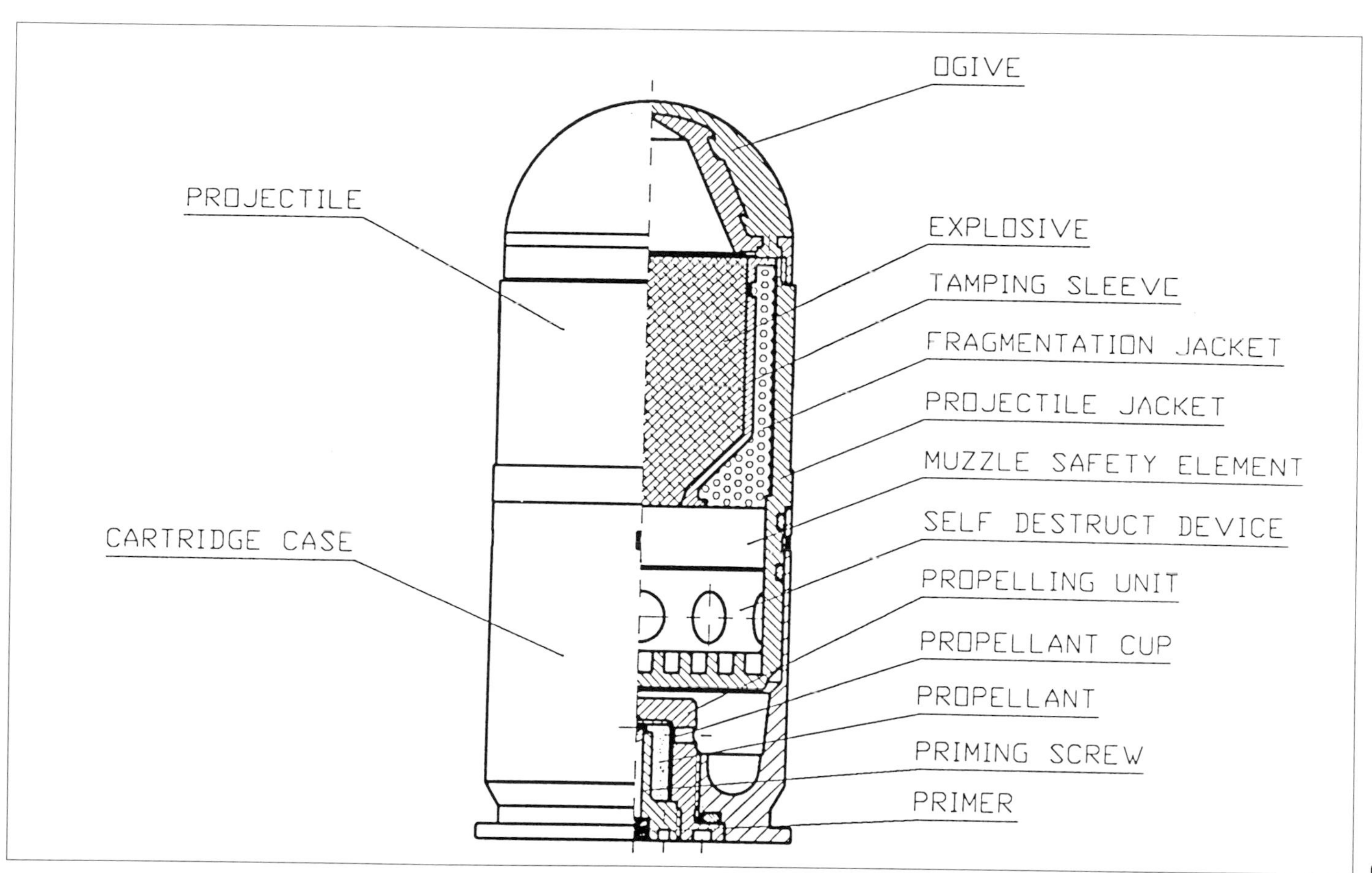
OGIVE
PROJECTILE
EXPLOSIVE
TAMPING SLEEVC
FRAGMENTATION JACKET
PROJECTILE JACKET
MUZZLE SAFETY ELEMENT
CARTRIDGE CASE
SELF DESTRUCT DEVICE
PROPELLING UNIT
PROPELLANT CUP
PROPELLANT
PRIMING SCREW
PRIMER

40mm Arges HEDP 92 Dual Purpose

Austria

This Austrian grenade looks similar to the HE 92 described previously, but functions in a different manner, offering both armour-piercing and anti-personnel effects. The cartridge is the same high-low pressure system as before and the grenade body has the same cylindrical shape with blunt nose, base fuze and self destroying element, but the inside of the grenade is somewhat different.

The same double skin is used, together with the same filling of about 1,000 2mm steel balls, but instead of a simple filling of high explosive, there is a shaped charge filling with a conical liner. This reduces the high explosive payload, so that the blast effect from the filling is slightly reduced, but this is compensated by the directional armour-piercing effect of the shaped charge. Moreover, as the shaped charge detonates to form the piercing jet, the detonation also fragments the grenade body and distributes the steel balls to give an anti-personnel effect against anyone in the vicinity of the target. It can be used in either role, against a hard target with nobody in the vicinity, or as an anti-personnel grenade with no hard target in view, if the need arises. This is not the weapon to deal with a main battle tank, but it can be effective against lightly armoured personnel carriers, soft vehicles and some forms of field fortification.

DATA
Calibre 40mm (1.57in)
Length, complete 4.0in (102mm)
Weight, complete 9.3oz (265g)
Weight, grenade 6.7oz (190g)
Grenade filling 0.7oz (20g) RDX
Fragments ca 1,000 2mm (0.078in balls
Muzzle velocity 250ft/sec (76m/sec)
Maximum range 437 yards (400m)
Penetration Over 1 inch (25mm) of steel armour plate.

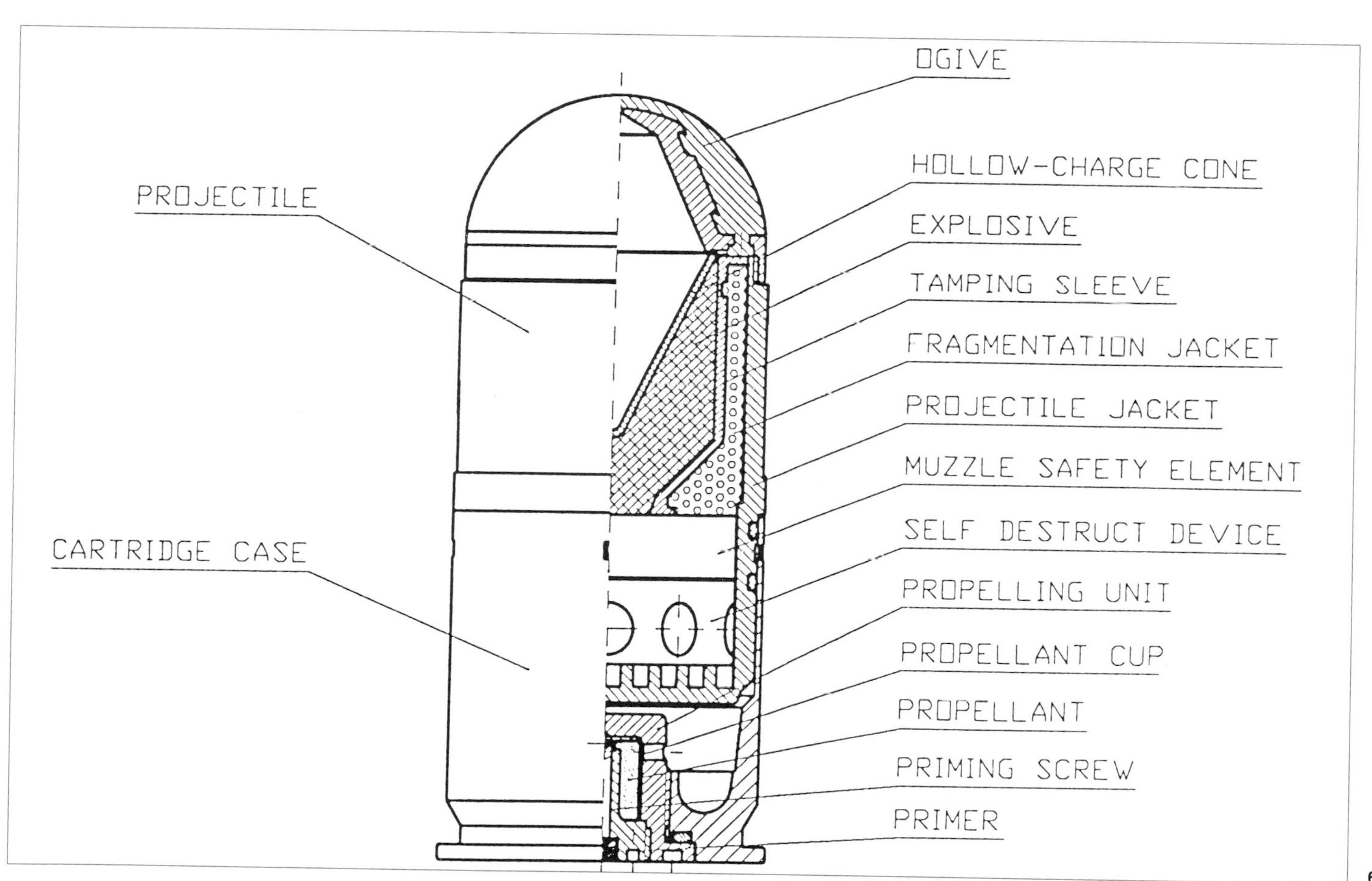
OGIVE
HOLLOW-CHARGE CONE
EXPLOSIVE
TAMPING SLEEVE
FRAGMENTATION JACKET
PROJECTILE JACKET
MUZZLE SAFETY ELEMENT
SELF DESTRUCT DEVICE
PROPELLING UNIT
PROPELLANT CUP
PROPELLANT
PRIMING SCREW
PRIMER
PROJECTILE
CARTRIDGE CASE

40mm Arges RP92 Smoke

This Austrian grenade is rather unusual in that it not only produces smoke, but it is also a first rate incendiary bomb, so that as well as screening activities it can also be used for 'smoking out' an enemy from a confined space such as a pillbox or a defended house.

The cartridge is to the same standard form as all short range 40mm rounds. The projectile is also, externally, of the same shape and size. It is, however filled completely (including the nose, which is usually left empty in explosive grenades) with red phosphorus and has a combined impact fuze and pyrotechnic delay element in the base.

After firing the impact element of the fuze will cause it to burst if it strikes any hard object and the red phosphorus mixture will be scattered around the point of burst where it will generate a cloud of dense smoke and will also ignite any combustible material upon which it lands. Should it not strike a sufficiently hard object to cause the fuze to operate, then at the end of about eight seconds from firing the pyrotechnic delay will burn through and explode the bursting charge, again scattering the red phosphorus and obtaining the desired effect.

DATA
Calibre 40mm (1.57in)
Length, complete 4.41in (112mm)
Weight, complete 7.76oz (220g)
Weight, grenade 5.11oz (145g)
Grenade filling 1.59oz (45g) red phosphorus
Muzzle velocity 250ft/sec (76m/sec)
Maximum range 437 yards (400m)

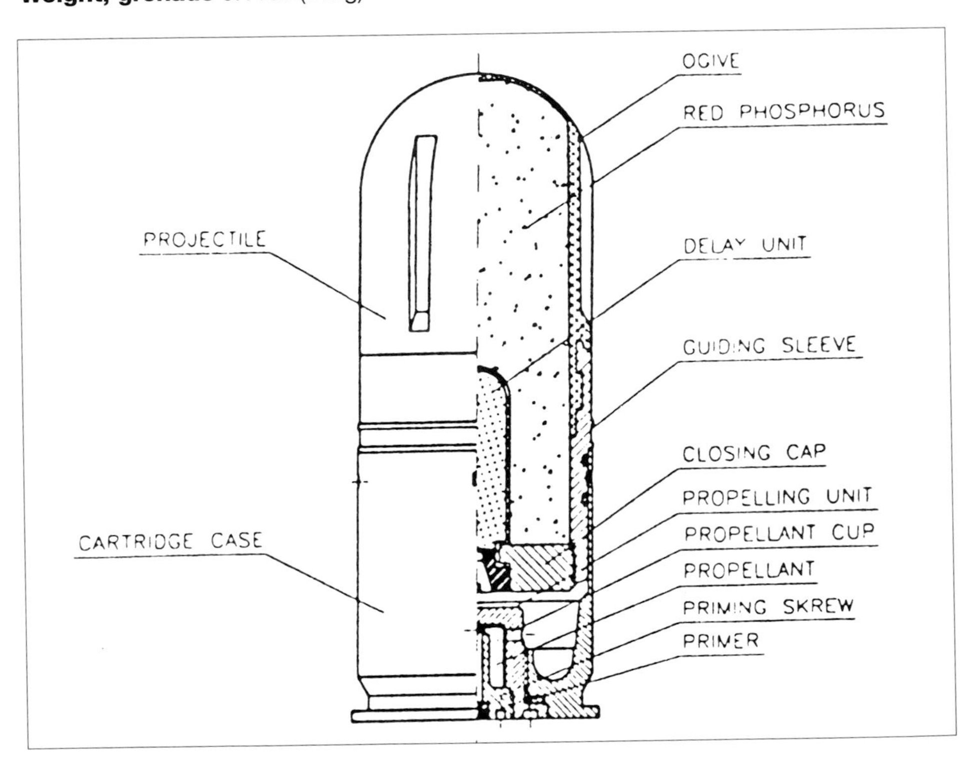

High Velocity HE M383

USA

With the arrival of the Mark 19 40mm machine gun – which, in spite of its name, is in fact an automatic grenade launcher – it was apparent that the velocity (and thus the recoil impulse) could be increased since the weapon was heavy and tripod mounted rather than light and hand fired. The obvious danger was that somebody would fire a long range cartridge from a hand held launcher and this was catered for by designing a new cartridge case 53mm long instead of the 46mm of the standard type. The case still uses the high-low pressure system, but the proportions of the two chambers have been changed, as has the diameter and number of holes through which the high pressure gas leaks into the low pressure chamber. The method of construction is also different, in that the spherical high pressure chamber is in two parts, the forward part formed inside the cartridge case and the rear part formed by a screwed in unit which also contains the ignition cap. The propelling charge is also larger than in the earlier designs of grenade.

The M383 round was developed in the late 1960s when the first automatic launchers were being tested as helicopter mounted weapons and has remained in service ever since. The grenade differs from earlier designs in having a curved base, more filling and an impact fuse inside the otherwise hollow nose. On firing, the propellant is ignited and burns at high pressure, leaking out into the cartridge case and propelling the grenade from the launcher. The difference lies in the quantity of gas produced and the much higher 'low' pressure, which gives the grenade a much higher velocity and longer range.

DATA
Calibre 40mm (1.57in)
Length, complete 4.41in (112mm)
Weight, complete 12.34oz (350g)
Weight, grenade ca 8.8oz (250g)
Grenade filling RDX/TNT
Muzzle velocity 794ft/sec (242m/sec)
Maximum range 2,405 yards (2,200m)

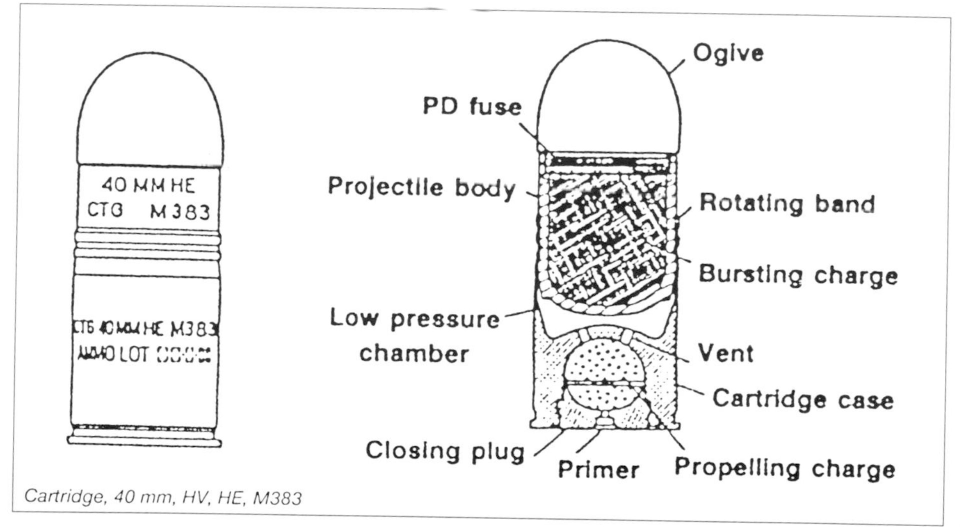

Cartridge, 40 mm, HV, HE, M383

40mm HE-PFF-T Anti-Personnel

This is a more modern example of the high-velocity type of 40mm grenade, developed by Diehl of Röthenbach for the German Army.

As the drawing shows, there are some differences from the original American pattern, notably in the design of the high-low pressure system compartment in the base of the cartridge case. The base of the case is solid and the propellant holder is screwed in from the front; it also has, in addition to gas vents, a hole in the middle through which the tracer unit of the grenade passes so that the flash of the propellant can ignite it.

PFF stands for pre-formed fragments and the shell of the grenade is lined firstly with a pre-notched coil of wire and, at the rear end, a number of steel balls. The pre-formed fragments are placed towards the rear of the grenade, behind the nose containing the fuze, so that on impact with the target the fragments are widely distributed. The body is then filled with explosive and the head carries an impact fuze with a pyrotechnic self destruction element.

On firing, the grenade is launched, the impact fuse is armed and the self destruction delay is ignited. If the grenade strikes a hard target then the fuze functions and detonates the explosive, scattering the fragments around the point of burst. Should the fuze fail to operate, as can happen if it lands in sand, snow, marshland or long grass, then after 18 seconds from firing, the pyrotechnic delay burns through and the explosive is detonated.

DATA
Calibre 40mm (1.57in)
Length, complete 4.41in (112mm)
Weight, complete 13.05 oz (370g)
Weight, grenade 8.64oz (245g)
Grenade filling 1.48oz (42g) RDX/TNT
Fragments ca 1,450
Muzzle velocity 791ft/sec (241m/sec)
Maximum range 2,187 yards (2,000m)

1) Primer;
2) Propelling charge;
3) Tracer;
4) Cartridge case;
5) Pre-formed fragments;
6) Driving band;
7) Explosive charge;
8) Grenade body;
9) Booster;
10) Self-destroying fuze.

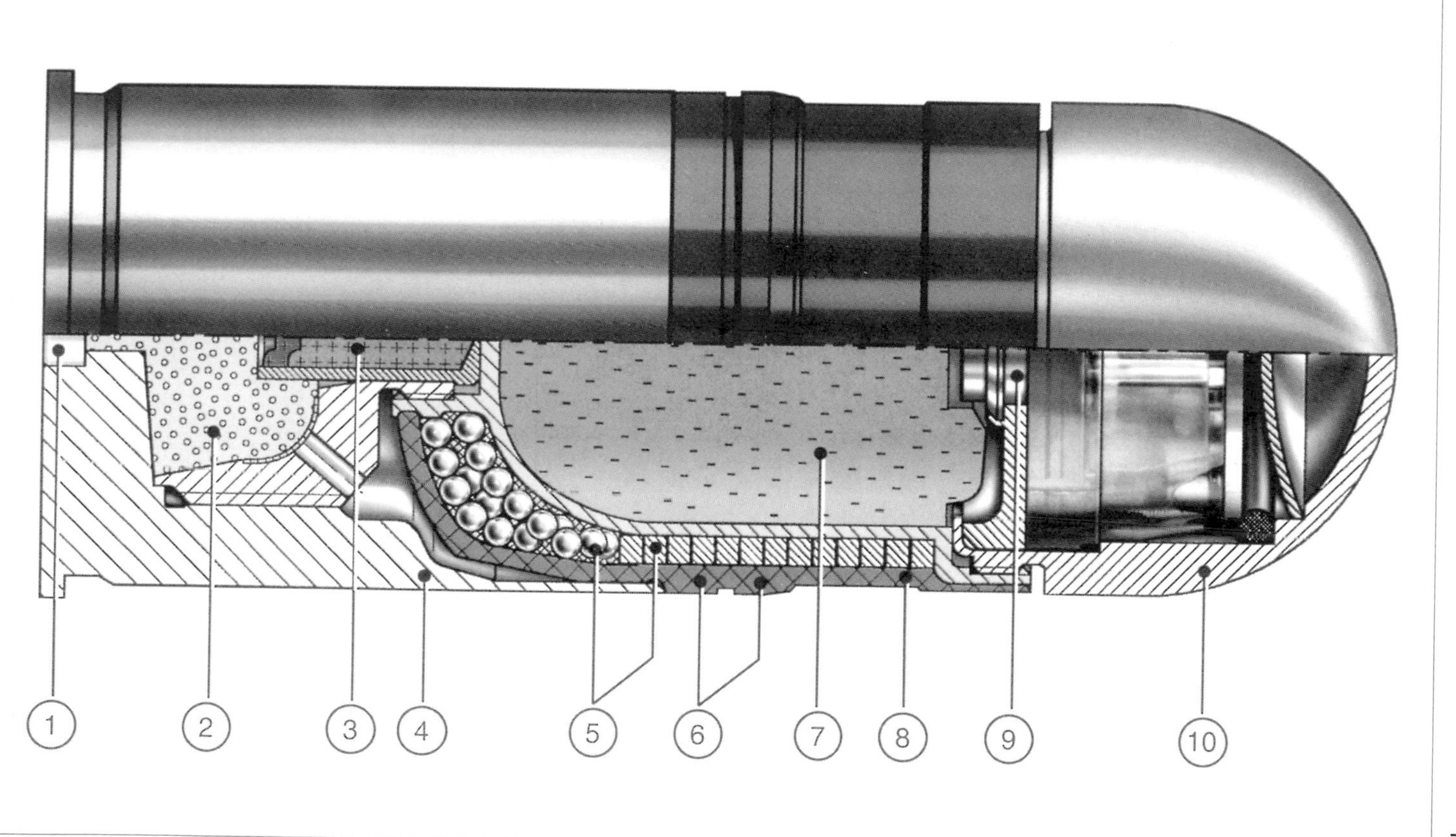
1
2
3
4
5
6
7
8
9
10

R1M1 Energa

The Energa grenade appeared shortly after World War Two from Switzerland; but 'the Swiss are only allowed to export munitions to people who don't want them', as one manufacturer once said and it therefore ended up in Belgium, from where it could be sold to anybody. For many years it was the only rifle grenade in British service and it was adopted by several countries in the 1955–75 period. Today it is less common and the example seen here is actually manufactured in South Africa and is somewhat improved over the earlier design.

The R1M1 is a shaped charge grenade, designed to be fired off the muzzle of more or less any 7.62mm rifle. The body is of steel and has a filling of RDX/Wax and a copper cone. The head of the grenade is merely a ballistic cap which ensures that the fuze will strike the target and initiate the shaped charge with a reasonable stand-off distance. The fuze is itself a tiny shaped charge; on striking the target this tiny charge fires a jet back down the axis of the bomb to strike a booster charge at the rear end of the main filling. This detonates and sets off the main filling which then collapses the copper cone to produce the penetrative jet.

The tail unit is a specially strengthened steel tube with a drum type fin assembly at the extreme end. Inside the tube is a small charge of smokeless powder.

To fire, the plastic protective cap is removed from the fuze and the grenade slipped over the rifle muzzle. A blank cartridge is loaded and fired; this generates a certain amount of pressure and it also ignites the smokeless charge inside the tail unit and this explodes to give even more propulsive gas, improving the velocity and range of the grenade.

DATA
Max diameter 2.87in (73mm)
Length, complete 14.56in (370mm)
Weight, complete 1.58lbs (720g)
Grenade filling ca 12oz (340g) RDX/Wax 97/3
Penetration 10in (250mm) steel armour
Muzzle velocity 203ft/sec (62m/sec)
Maximum range 375 yds/m

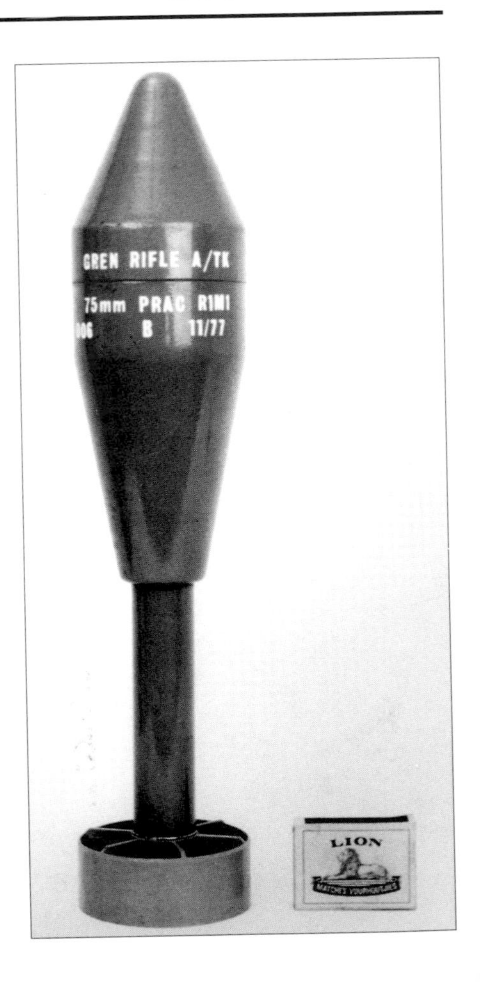

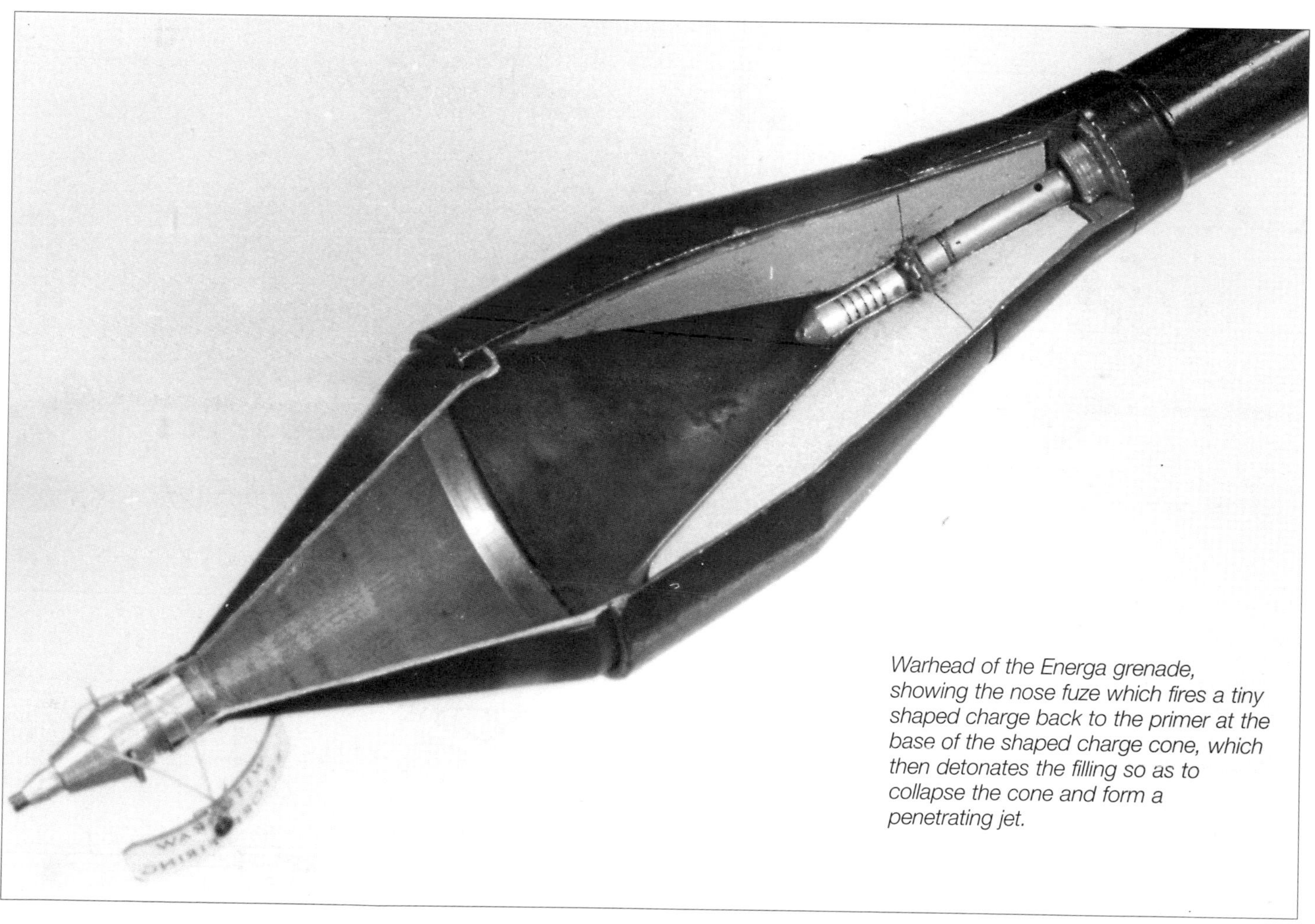

Warhead of the Energa grenade, showing the nose fuze which fires a tiny shaped charge back to the primer at the base of the shaped charge cone, which then detonates the filling so as to collapse the cone and form a penetrating jet.

BT/AT 52

Israel

This is an example of the 'bullet trap' type of grenade, a design which has become very popular and widespread over the past 20 years. The object is to remove the need to have a special blank cartridge for launching a grenade. This was of little account when bolt action rifles were standard, but it became something of a nuisance when semi-automatic weapons were adopted. It then became necessary to remove the magazine, empty the chamber, hand load a blank cartridge, turn off the gas regulator which drove the semi-automatic mechanism, load the grenade, fire it, eject the spent case, replace the magazine and reload the rifle. In the face of a sudden attack, this could be a fatal delay.

The BT/AT 52 has a steel tail unit, inside which are a series of steel discs and a hardened steel bullet deflector which ensures that if, by some freak, the trap does not stop the bullet it will be deflected out through the side of the tail tube and will not strike the actual warhead and perhaps detonate it. The steel discs are carefully designed so as to collapse in succession as the bullet strikes them, each one slowing the bullet until the energy is entirely dissipated. The trap can deal with either lead core or steel core bullets, so that any cartridge loaded in the rifle chamber, ball, AP or tracer, or even a grenade blank cartridge, will serve to launch the grenade. The launch impulse is supplied by a combination of the energy delivered by the bullet and the propulsive effect of the propellant gas behind the bullet.

The warhead is a conventional shaped charge with an impact fuze capable of detonating the charge at angles of impact as low as 15 degrees.

DATA
Diameter 1.96in (50mm)
Length, complete 15.75in (400mm)
Weight, complete 1lb 2oz (510g)
Grenade filling RDX/TNT
Maximum range ca 300yds/m

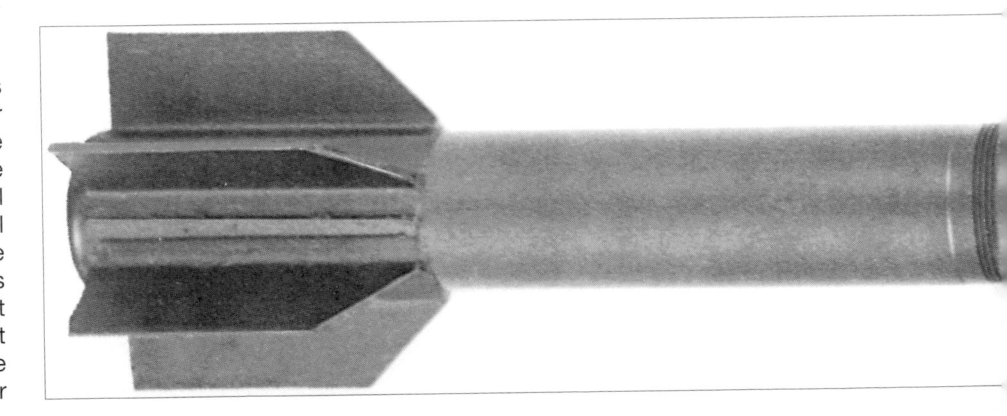

(Right) Examples of the bullet trap system; top, the trap with steel energy absorbing discs before use; bottom, after firing – the bullet is scarcely recognisable, beneath the third disc.

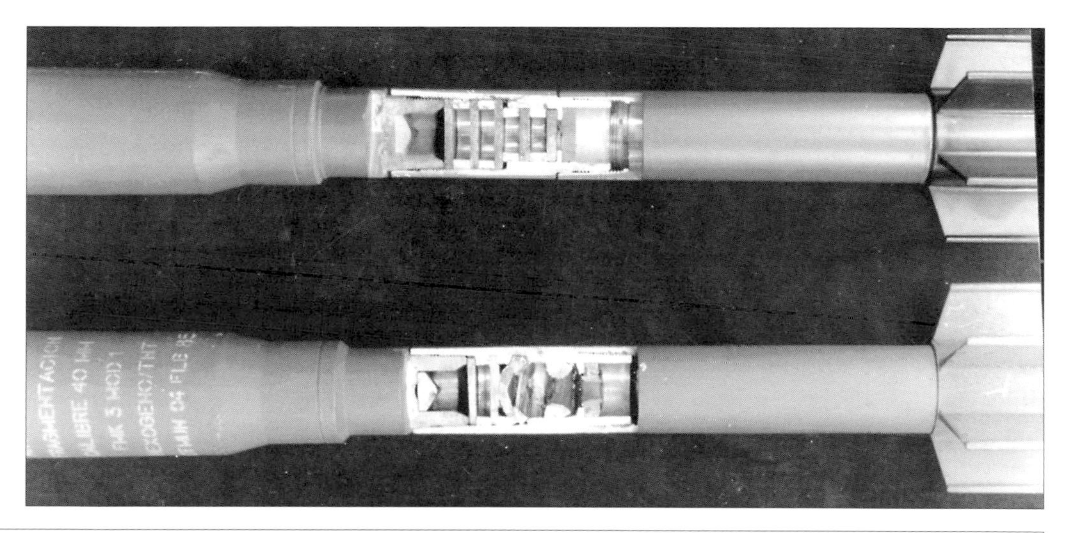

FN Bullet-Thru™

The alternative to the bullet trap is this 'bullet through' system which allows the bullet to pass completely through the grenade. For training it can be fired with a grenade blank cartridge, so obviating the need for the large safety area demanded by earlier bullet through systems.

The FN grenade consists of two parts, head and tail. The tail is telescoped into the head for shipping and transportation, to save space. When telescoped the detonator is out of alignment with the firing pin and a safety shutter separates the detonator from the explosive. To fire the grenade, the tail is pulled out from the head; the detonator (in the head) is now separated from the firing pin (in the tail) and the tail also carries the fragmenting sleeve, now removed from the explosive charge in the head. The grenade is placed on the muzzle and fired by whatever cartridge is in the rifle. The bullet passes through a polycarbonate plug in the grenade and off into the distance; there is sufficient energy in the bullet and gas to launch the grenade from the rifle and as it leaves the muzzle a spring causes the tail unit to retract into the grenade body once more. While doing this, the tail rotates so as to bring the firing pin into line with the detonator and as the two parts come into position, so the fragmentation sleeve slides around the explosive charge. The fuze arms during the first few metres of flight and after that it will detonate the explosive filling as soon as it strikes the target.

As well as this anti-personnel grenade, FN have designed shaped charge, illuminating, smoke and training grenades using the same mechanical principles. The grenade is also manufactured under licence in Indonesia. An advantage of this design is that for training purposes the grenade can also be launched by a blank cartridge.

DATA
Diameter 1.45in (37mm)
Length, collapsed 7.44in (189mm
Length, extended 11.41in (290mm)
Weight, complete 11.3oz (320g)
Muzzle velocity 262ft/sec (80 m/sec) (7.62mm rifle)
Maximum range 400yds/m (7.62mm rifle)
Lethal radius 10yds/m

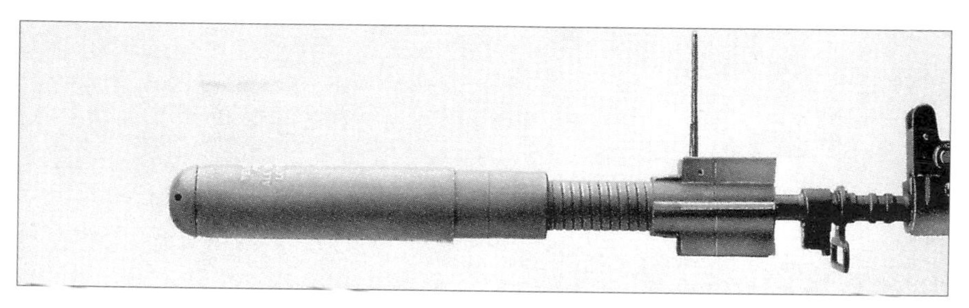

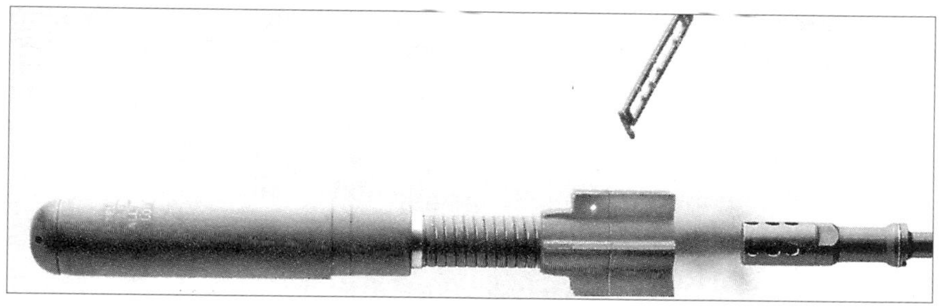

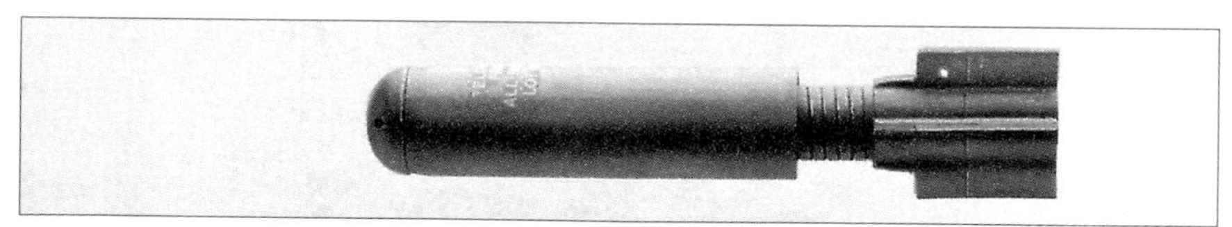

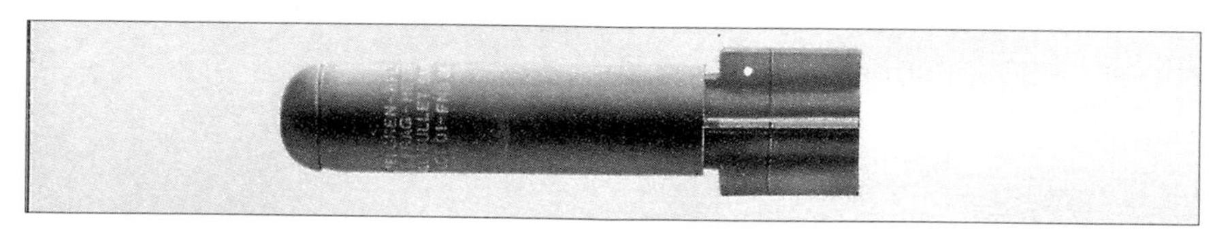

(Left) The flight of the Bullet-Thru™ grenade (top to bottom): fully extended and ready to fire; firing, contraction and arming; and the final configuration.

The Polyvalent Grenade

This was developed in the 1970s by the Losfeld company of France, and is an interesting example of making one basic grenade fit into three roles. It was adopted by the French and several other armies and may well still be in use with some of them.

The Polyvalent grenade could be adjusted to use in three forms: an anti-personnel rifle grenade, a defensive hand grenade or an offensive hand grenade. The change of tasks could be made very quickly, since the grenade was a combination of three elements; the explosive body, a fragmentation sleeve and the finned tail assembly. The fuze permitted selection of three different methods of functioning: on impact, after five seconds of delay, or impact and the delay element should impact alone fail. There were also three types of tail unit; the F1 could only be used with a grenade launching blank cartridge, the F556 was a bullet trap type suitable for use with 5.56mm rifles and the F762 was a bullet trap of stronger construction for use with 7.62mm rifles.

To use as an offensive grenade the explosive body was fitted with the fuze, 'impact' selected and the grenade thrown. As it left the hand, so a safety tape unwound and pulled out the safety pin from the fuze, arming it ready to function on impact. To use as a defensive grenade the fragmentation sleeve was slipped over the explosive body and locked in place, after which the fuze was set and the grenade thrown as before.

To use as a rifle grenade the fragmentation sleeve was fitted, an appropriate tail unit was screwed to the base of the grenade, the fuze set to the selected function and the grenade fired from the rifle.

DATA

Length, complete 14.17in (360mm) with tail unit F762

Weight, complete 1.14lbs (520g) with tail unit F762

Muzzle velocity ca 310ft/sec (95m/sec)

Maximum range ca 225yds/m

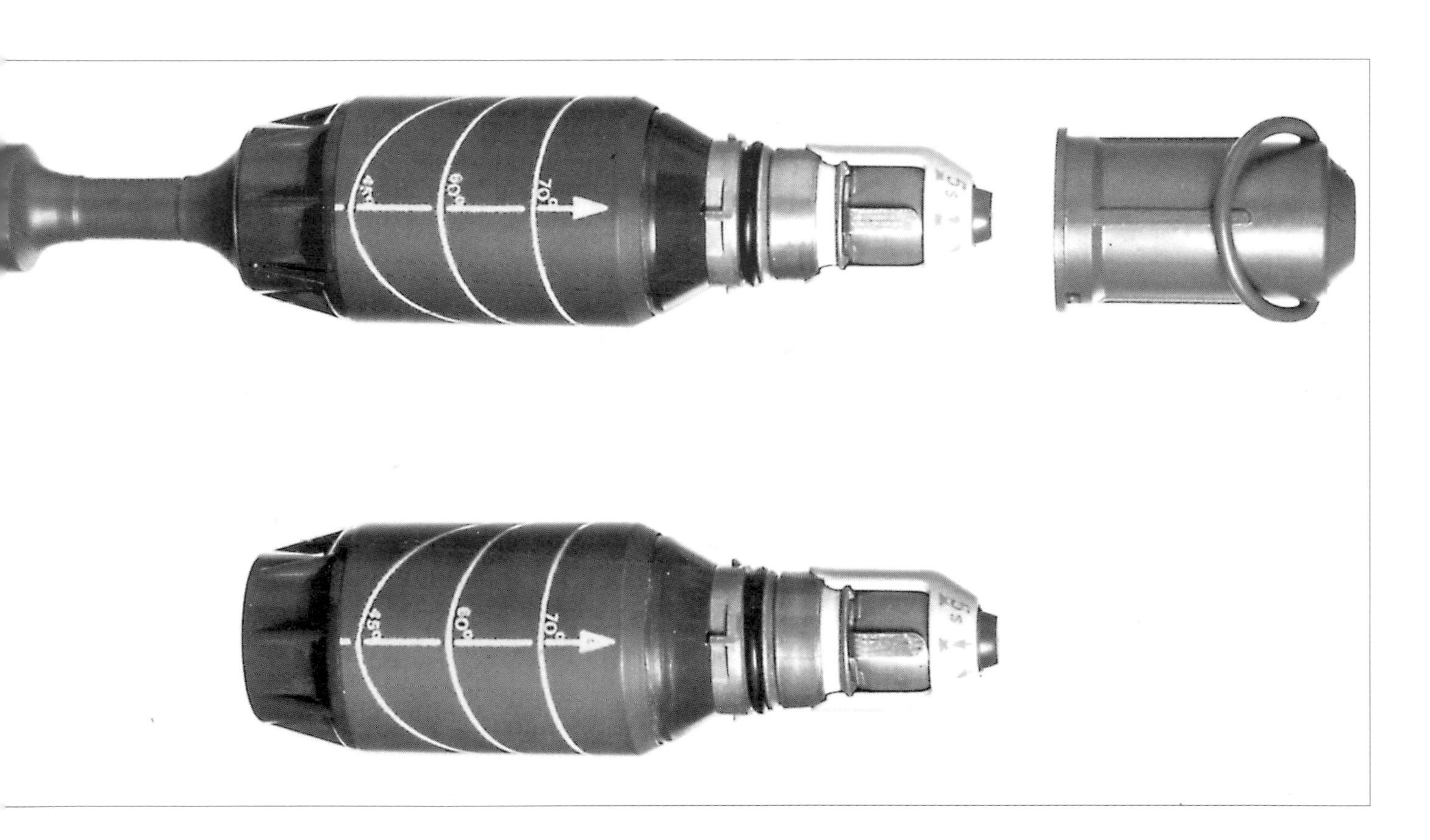
45°
60°
70°
45°
60°
70°

Mortar Ammunition

Originally, the mortar was an inexpensive steel tube with a firing pin on the bottom end, balanced on an inexpensive sort of tripod, and firing cheap ammunition. The mortar was termed an 'area weapon' and there were two kinds of bomb, high explosive and not. Fire control was a soldier with a pair of binoculars and a map.

After some time, it was realised that if the mortar was more accurate, longer ranging and had some better bombs, it might cease to be an 'area weapon' and become more cost effective – more hits meant fewer bombs wasted. Around 1960 the mortar revolution started and today's mortar is far more accurate, far more consistent and hence far more dangerous to an enemy, than ever before. It uses electronic fuzes and computerised fire control. But it is still, compared to most other heavy weapons, simple to operate, although the crew need to be computer literate.

But without improvements in the ammunition, the improvements in the mortar would never have come about; by lifting the ammunition out of the `area weapon' category, the mortar designers had an incentive to improve the weapon, though in fact the improvements to the actual mortar amount to little more than building to an improved set of tolerances and with a better degree of finish. Essentially, it is still a steel tube balanced on a bipod and baseplate.

The problem with the mortar was that one had to drop the bomb down the barrel to load it and this had to hit the firing pin in the bottom of the barrel with sufficient force to fire the cap and ignite the propellant. Some mortars had a trip operated firing pin which reduced the need for a fast descent of the loaded bomb, but the principles were the same and so was the underlying problem. This was simply that if the bomb was made to have too good a fit in the bore, the air trapped underneath the bomb slowed down its descent so that it didn't hit the pin hard enough, resulting in a slow rate of fire from a weapon which generally relied upon having a fast rate of fire. If the bomb was made to have a loose fit in the bore so that it loaded easily and fell quickly, then when the propellant exploded and generated gas, far too much of the gas escaped around the bomb and, because the bomb tended to bounce off the sides of the barrel on its way out, the accuracy was very much reduced. The difference in diameter between the exterior of the bomb and the interior of the barrel, technically called 'windage', became a critical design factor; some bomb designers put multiple grooves in the widest part of the bomb so that the gases would swirl and become trapped there and act as a self forming seal for the duration of the bomb's travel up the barrel. It worked after a fashion, but did nothing to improve accuracy.

One way to overcome this problem was to rifle the mortar barrel, put a soft

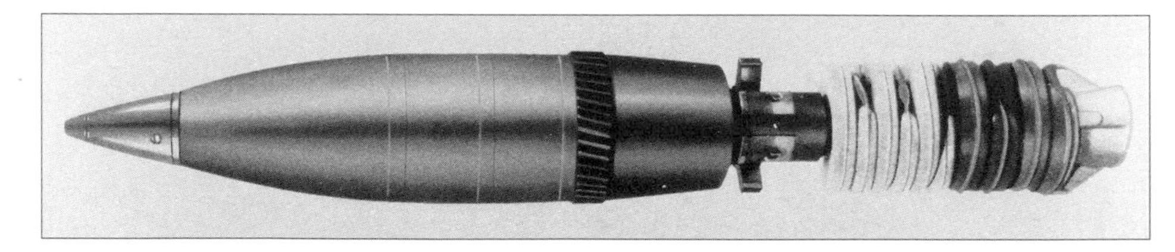

A bomb for the Brandt 120mm rifled mortar; note the pre-engraved driving band which must be fitted into the rifling to load

Loading the 120mm rifled mortar.

metal driving band on the bomb and pre-cut grooves in the driving band so that loading became a question of fitting the grooves in the bomb over the lands of the rifling before letting the bomb slide down the bore. This was the approach of Brandt, the French designer. Another approach was due to an Australian officer, who placed a soft metal saucer at the bottom of a bomb which resembled an artillery shell. Beneath the saucer was a round steel plate and behind this was a tube containing the propelling charge. The entire bomb, saucer and plate had plenty of windage and slid down the bore very easily. When the charge hit the pin and exploded, the blast drove the steel plate forward and flattened the soft metal saucer, forcing it out into the rifling. The American army adopted this as their 4.2 inch chemical mortar in the 1930s and it worked very well. Both these ideas however were essentially reversions to the 1870s technology of the rifled muzzle loading gun. For the most part, it was a case of `windage or nothing'.

At the end of World War Two all mortar bombs, with a few minor exceptions, were to the same general design: a cast iron tear drop shaped bomb with a tubular tail boom bearing a set of stabilising fins. The tail boom contained the 'primary cartridge', almost always a reworked shotgun cartridge containing a charge of fast burning smokeless powder. Clipped, tied, wedged or otherwise distributed around the tail boom and fins, would be the 'secondary charges', capsules, bags or simple sheets of propellant which could be adjusted to provide different charge zones and which were ignited by the primary cartridge. The bomb was usually filled with the cheapest possible explosive – Amatol or some similar compound which would break the cast iron into suitable fragments and generate sufficient blast to damage material targets. The explosive charge was detonated by an impact fuze. The only other types of bomb to be found were white phosphorus smoke, similar in shape and size to the high explosive bomb and illuminating bombs containing a white flare and a parachute.

Notable departures from this general form were the American 4.2 inch rifled mortar which used the same types of bomb, but dispensed with fins and used the expanding driving band referred to above, and the British 4.2 inch which used a discarding sabot tubular bomb to deliver smoke canisters, though its HE bomb was of the usual type.

Throughout the war years the British army had been conscious of the deficiencies of their standard 3 inch mortar, which was consistently outperformed by the German 81mm equivalent. The 4.2 inch was also outclassed by the 120mm weapon, which the Germans had adopted from the Russians. During the Korean War the 4.2 inch was used more widely, due to the mountainous terrain. The weapon exhibited alarming defects, notably a tendency to split the tail boom on firing and thus drop the bomb short; this was probably due to the extreme cold affecting the welding of the fins to the tail boom. In the late 1950s research into a new and improved mortar began, resulting in the 81mm mortar.

The mortar itself was as before, but using stronger steel for the barrel and a generally higher quality of construction; the innovation lay in the ammunition. Instead of a roughly cast bomb with minimal finish, the 81mm L16 bomb was of high grade austenitic steel, carefully machined to exact dimensions. It was smoothly tapered and had a tail boom with fins which had been machined from solid aluminium alloy and, again, carefully dimensioned and finished. The most startling feature was a groove running round the body of the bomb at its widest part, into which fitted a ring of Makrolon plastic material, carefully shaped. A new impact fuze, shaped to match the taper of the bomb nose, went into the head end and the usual sort of primary cartridge went into the tail boom, around which were clipped a number of secondary charges in horse shoe shaped plastic containers.

The bomb was drop loaded in the traditional manner; the windage was reduced to the minimum compatible with allowing the bomb to slide down the bore with sufficient force to fire the cartridge when it hit the firing pin. Here the innovative design came into play.

The Makrolon ring around the bomb was, in effect, a skirt; as the charge exploded and the propelling gas rushed up the sides of the bomb, so the pressure forced this skirt out into tight contact with the interior of the barrel, wedging the inner part of the skirt into the groove in the bomb and making a most effective gas seal. It also centered the body of the bomb in the bore, the tips of the tail fins also being in contact and thus the bomb went up the bore as nearly as possible aligned with the bore axis. It left the bore with a slight yawing effect, as does every fin stabilised mortar bomb, but soon followed a very accurate and consistent trajectory. Due to this efficient gas seal and efficient centering, the bomb had greater range and far greater accuracy than any previous mortar bomb in history. It was no longer an 'area weapon'; the bomb could be placed practically anywhere as required.

Once the British 81mm mortar had shown the way, there was no lack of followers. The US Army looked at the situation and, surprisingly, chose to build the British mortar under licence. They added a small but significant refinement on the bomb, however, canting the tail fins very slightly so that the bomb rolled as it flew through the air, thus imparting even greater stability and consistency. Other countries adopted the same, or similar, form of sealing ring around the bomb and extended the idea into other calibres.

The British have always regarded the mortar as an infantryman's weapon, for his own immediate support. For anything more serious artillery is readily available. But many other countries have a heavy mortar which, whilst capable of dealing with close targets, has a near artillery range capability and can give quite heavy support to an infantry attack. The favoured calibre for this is 120mm and there is a wide selection of mortars available. It soon became obvious that the same technology would rapidly find its way into the 120mm field and bestow its benefits. The French Hotchkiss-Brandt rifled 120mm mortar had hitherto been the undisputed champion in this calibre, since its rifled system gave better stability and gas sealing, hence more range and better accuracy than comparable smooth bores, but the new gas sealing bombs were soon in contention. Brandt produced a rocket assisted bomb which gave quite a surprising increase in range, though not in accuracy.

In the 1970s the artillery ammunition manufacturers began investigating a variety of 'Improved Conventional Munitions' (ICMs – which we will examine later) which, in broad terms, means filling shells with mines and bomblets and scattering them over the target area. This also interested the mortar bomb makers, but there seemed to be little prospect of placing much of a payload into a mortar bomb. The object of these ICMs was the attack of tanks at ranges well beyond the range of the tank's own weapons before they could begin to exert any influence on the battlefield. Indeed, if the ICMs worked, the tanks wouldn't even reach the battlefield. Interest was aroused in the possibility of a dedicated anti-armour mortar bomb being used by the infantryman to keep enemy tanks at bay.

Even the most dedicated and enthusiastic mortarman had to admit that his chances of a direct hit on a tank turret at, say, 5,000 yards, were remote, so simply putting a shaped charge into an ordinary mortar bomb was obviously a waste of effort. Thus three different companies in three different countries began studying the possibility of guiding a mortar bomb to its target. Sweden and Germany developed bombs for their 120mm mortars, while Britain was somewhat handicapped by having to fit everything into its 81mm calibre. All three worked in much the same way; they were fired like an ordinary bomb, discarded their tail units shortly after leaving the muzzle, extended wings or fins and on their downward trajectory scanned the ground below, using various types of sensor, to detect a tank target. Having found one they then steered themselves to impact on the top surface where the armour was thinnest. By the middle 1980s all had proved successful in trials and thereafter the question was one of engineering them into a produceable form. Unfortunately, the devices work, but at a staggering price. Whether or not they achieve any significant military acceptance as a result remains to be seen.

Another approach, pioneered in the USA in the 1980s, was the FOMP: Fiber Optic-guided Mortar Projectile. Here the nose of the bomb held a video camera which fed its picture back to an operator near the mortar. He could also transmit steering commands down the fibre optic which trailed behind the bomb, so that he could actually steer the bomb on to the target of his choice. Again, it has been shown to work, but budgetary considerations appear to be delaying its advancement.

By the mid 1980s, technology had moved forward again and now there appeared the first of several mortar bombs carrying sub-projectiles. These take the form of small shaped charge bombs wrapped in fragmentation sleeves. The bomb is fitted with a time fuze which causes it to split open over the target and release the bomblets so that the scatter over an area of several square yards. Each bomblet is capable of piercing light armour – up to about 50mm thick – and also shatters its fragmentation sleeve so as to generate a cloud of lethal anti-personnel fragments in the area. So if it fails to hit a tank it can still do some harm to any personnel standing around. Bombs of this type are now in service in Greece and Spain.

In years gone by, the only fuze in regular use in mortars was an impact fuse, as time fuzes demanded firing tables and calculations to get the burst at the right height above the target. Modern electronic fuzes, associated with computerised fire control systems, mean that the calculation and setting of fuzes takes mere seconds and thus it becomes feasible to employ time fuzes in roles such as the sub-projectile bomb. Moreover, the reduced size and cost of modern proximity fuzes makes them an economic proposition for use with mortars and they do not require any calculations or setting; they will detonate themselves at the correct lethal height above a target without any human intervention.

From this brief exposition, it can be appreciated that today's mortarman has a considerably more effective weapon system at his disposal and most of that efficiency is due to improvements in ammunition. The section which follows will examine some of this in more detail, starting with some elderly equipment as a basis for comparison, looking at the equipment which is currently in use and looking forward to what is undergoing evaluation.

Modern mortarmen; a Norwegian mortar team using computerised fire control.

82mm HE Bomb

Czech Republic

This is a text book example of the old style of cheaply manufactured bomb with which virtually every army was equipped until the 1950s. The bomb body is of cast iron and is left exactly as it came from the mould, apart from machining the bourrelet, the band around the widest part of the body, to the proper dimension. The tail end of the bomb is formed into a spigot and threaded and the drawn steel tail tube is screwed on. The stamped steel fins are welded to the tail tube with two spot welds between each pair of fins. A primary cartridge went into the tail tube and the secondary charge was wrapped around the tube in a muslin bag and held by a piece of wire. The bomb was then filled with amatol, a mixture of 40 percent TNT and 60 percent ammonium nitrate and a simple impact fuse screwed it to the nose.

The markings on this bomb indicate that it was made in 1952, before the revolution in mortar ammunition design began, and it probably cost less than £1 to manufacture and fill.

DATA
Calibre 3.23in (82mm)
Length overall 13in (330mm)
Weight as fired 6.83lb (3.10kg)
Number of charges Primary + 3
Muzzle velocity Approx 655ft/sec (200m/sec)
Range Approx 2,185 yards (2,000 metres)

82mm HE Bomb O-832A

Bulgaria

This is a slightly more advanced design, probably dating from the early 1940s and copied from a Soviet army bomb. The shape is still the traditional tear drop and the body is of cast ferro-steel, but the bourrelet has been carefully machined into a number of grooves. These act as a gas trap, building up a pressure barrier between the bomb and the barrel of the mortar, which prevents more of the propelling gas getting past and wasting its energy. The material of the body is somewhat tougher than cast iron and thus the filling is pure TNT which will break the ferro-steel into suitably sized anti-personnel fragments. The rear end of the bomb is hollowed out and threaded and the tail tube is a machine steel casting, with pairs of fins spot welded on, as before. A primary cartridge can be seen inside the tail boom and the secondary charges would have been in horse shoe shaped plastic containers which clipped around the tail in front of the fins.

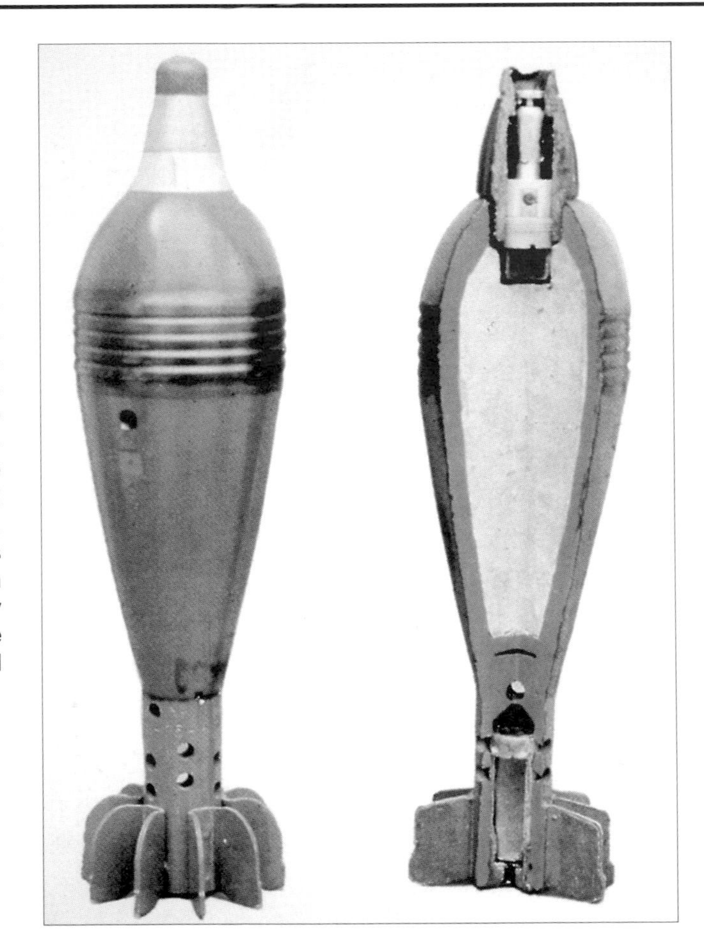

DATA
Calibre 3.23in (82mm)
Length overall 13.0in (330mm)
Weight as fired 7lb (3.18kg)
Number of charges Primary + 3
Muzzle velocity 738ft/sec (225m/sec)
Range 3,335 yards (3,050m)

120mm HE Bomb OF-843A

Bulgaria

Another Bulgarian copy of a wartime Soviet design, this time for the 120mm mortar which the Soviets pioneered and which was widely copied. It has a modified tear drop shape, still of the same general form, but more gracefully tapered, because the large calibre gives the designer more room to move. The construction is similar; the body is of cast ferro-steel with five gas check grooves machined into the bourrelet. The remainder of the body has been machined sufficiently to give a smooth surface so as not to generate drag in flight, but without unduly worrying about close tolerances. The nose has a fairly large filling hole which is closed by a bush threaded to take the impact fuze. The filling is 1.6kg of TD 50 which is probably a 50/50 Amatol mixture; the Russian equivalent is filled with 2.6kg of TNT.

The rear end of the bomb is hollowed out and threaded to take the machined steel tail boom with the usual four pairs of spot welded fins. A primary cartridge fits into the tail boom and six secondary charges can be clipped around the tail ahead of the fins.

DATA
Calibre 4.7in (120mm)
Length overall 26.18in (665mm)
Weight as fired 35.2lb (16.0kg)
Number of charges Primary + 6
Muzzle velocity 915ft/sec (279m/sec)
Range 6,365 yards (5,820 metres)

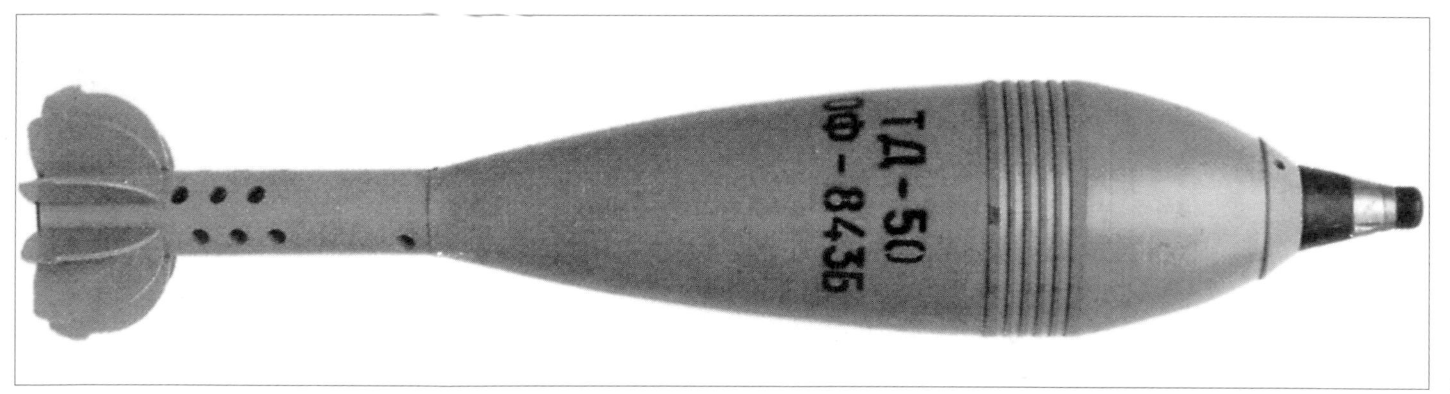

60mm HE Bomb NR431A1 — Portugal

In the late 1960s the Belgian company Poudres Réunies Belgique (PRB) took a hard look at the mortar bomb and decided that one of the basic problems was inefficiency due to poor fragmentation, particularly in small calibres. So they came up with this ingenious design, which they sold with some success. PRB are now out of business, but the bomb is still manufactured in Portugal for the Portuguese Army and for export.

As can be seen from the drawing, the actual bomb body is of fairly thin cast steel. It is roughly finished, except for the bourrelet which is finished to dimension and has two shallow gas check grooves. A spigot is welded to the bottom of the bomb and the machined alloy tail unit is press fitted on to this spigot and pinned in place. The tail boom and fins are all in one piece, machined from a casting. Inside the body is a patented 'fragmentation sleeve' which is simply a coil of square section notched steel, shaped on a former so that the coil can be squeezed to go into the wide opening at the front end of the body and then expands outwards to lie snugly against the bomb walls. A nose bush is then screwed in, locking the fragmentation sleeve in place and the bomb is then filled up with 150 grams of TNT. An impact fuze goes into the nose, a primary cartridge and four secondaries on to the tail unit and the bomb is complete. On striking the target the TNT detonates and shatters the bomb body; it also shatters the fragmentation sleeve into thousands of small but lethal fragments which can be effective up to about 15 metres from the point of burst, which is very good efficiency indeed for a 60mm bomb.

DATA
Calibre 2.36in (60mm)
Length overall 10.03in (255mm)
Weight as fired 3.0lb (1.36kg)
Number of charges Primary + 4
Muzzle velocity 580ft/sec (177m/sec)
Range 2,300 yards (2,100 metres)

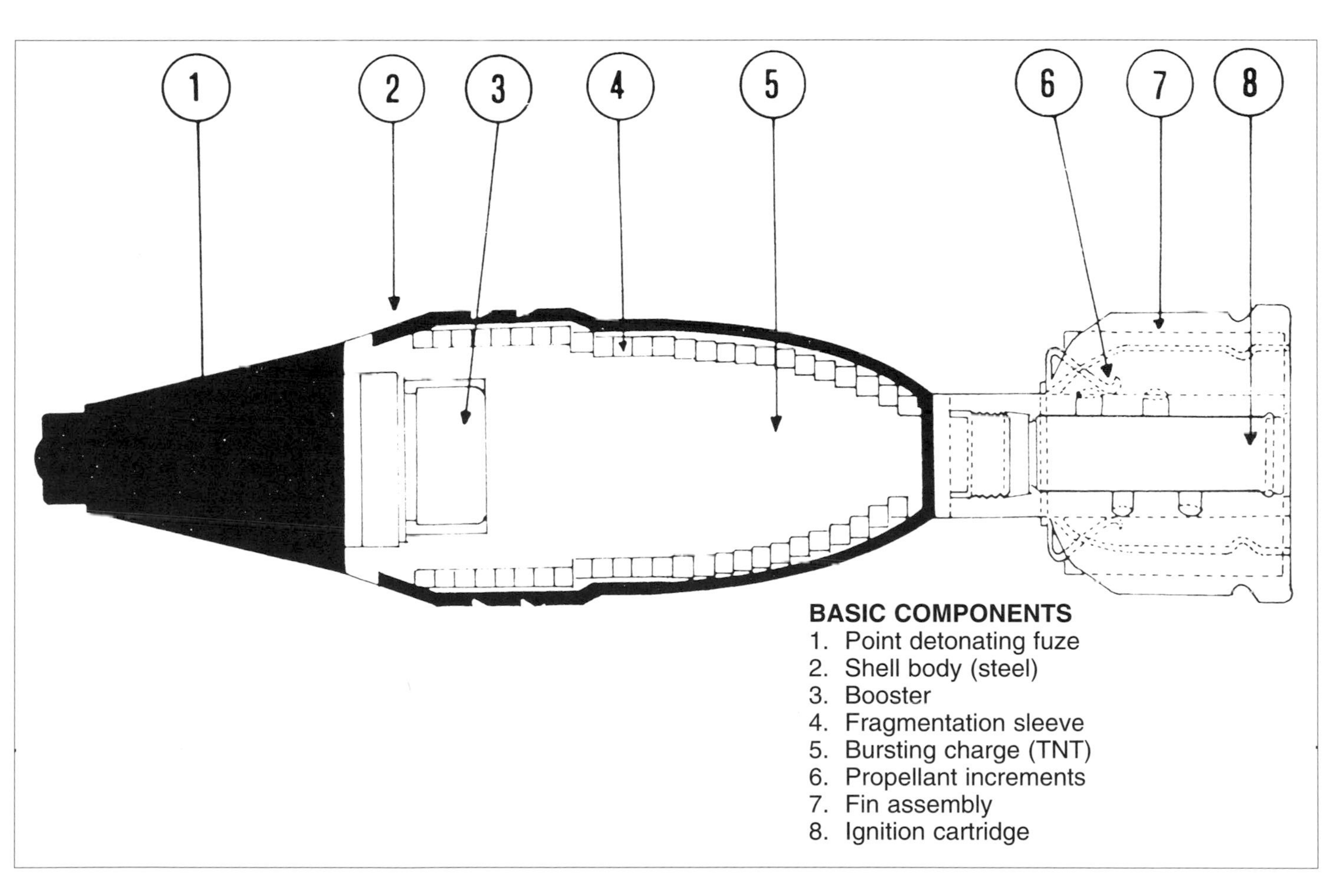
1
2
3
4
5
6
7
8
BASIC COMPONENTS
1. Point detonating fuze
2. Shell body (steel)
3. Booster
4. Fragmentation sleeve
5. Bursting charge (TNT)
6. Propellant increments
7. Fin assembly
8. Ignition cartridge

81mm HE Long Range Bomb M64 — Israel

Once the British 81mm bomb became public knowledge, there was a rush to adopt the same design and this Israeli bomb is a near perfect copy of the British 81mm L36 design. The most obvious difference lies in the secondary cartridge; in the British design they are six celluloid horse shoes filled with propellant, while in this Israeli design they are seven fabric covered horse shoes; both types fit tightly around the tail boom.

The bomb body is of spheroidal graphited cast iron, which produces the optimum size and shape of fragment from the detonation of the 740 gram TNT filling. The second difference from the British bomb is that the nose is threaded to accept a fuze; in the British design there is a larger opening at the nose, filled with a threaded bush into which the fuze fits. There is no particular virtue in either design, it is simply a case of building the bomb to suit one's particular methods of manufacture. The tail boom and fins are a one piece unit, machined from a light alloy extrusion, and screw into a threaded recess in the tail end of the bomb body. The fuze is the DM111, made by Junghans of Germany, and has a safety wire which must be removed before firing and a short delay element which can be set by rotating the prominent screw head seen on the side of the fuze. This gives a delay of 0.06 of a second between impact and detonation, sufficient for the bomb to pass through light cover (such as a roof) before bursting.

DATA
Calibre 3.19in (81mm)
Length overall 18.7in (475mm)
Weight as fired 8.6 lbs (3.90kg)
Number of charges Primary + 7
Muzzle velocity 935 ft/sec (285 m/sec)
Range 5,100 yards (4,660 metres)

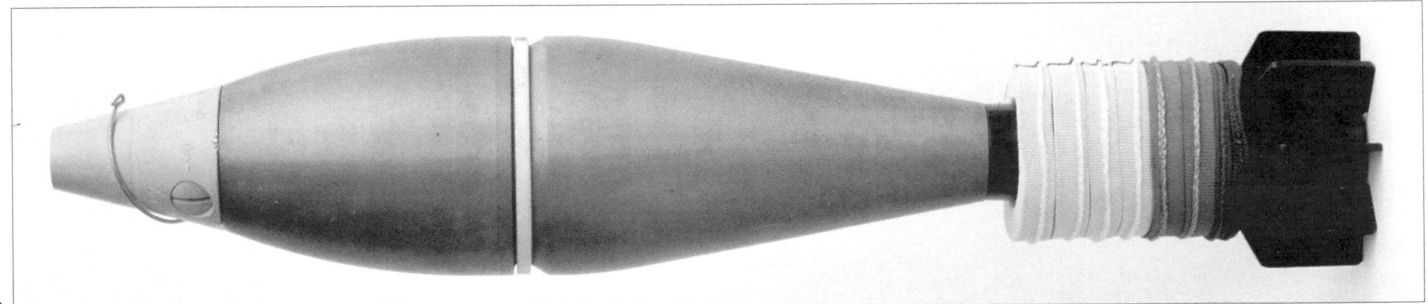

81mm HE Bomb Model 84AE — Spain

This slender and elegant bomb was developed by the Spanish company ECIA (Esperanza y Cia) who have been mortar specialists since the 1920s; they designed the 2 inch mortar which was used by the British Army during World War Two and which is still used in India and Pakistan.

It can be seen that it is simply an extension of the British L36 design, longer and more tapering, but adhering to the same system of using a plastic sealing ring around the bourrelet to provide an efficient gas seal. The body is forged from ferritic and pearlitic steel and loaded with just over 1kg of TNT, giving a very good effect at the target. The tail boom is machined from alloy, has a primary cartridge inside and takes up to six secondaries clipped around the boom ahead of the fins. An impact fuze is fitted in the nose, with a safety pin which has to be removed before loading.

The Spanish have two 81mm mortars in service in which this bomb can be used. One has a 1.15 metre barrel and the bomb has a maximum range of 6,200 metres; the other has a 1.45m barrel and gives the bomb a range of 6,900 metres, which is quite remarkable for an 81mm weapon.

DATA
Calibre 3.19in (81mm)
Length overall 21.2in (539mm)
Weight as fired 9.92lb (4.50kg)
Number of charges Primary + 6
Muzzle velocity 1,082ft/sec (330m/sec)
Range 7,545 yards (6,900 metres)

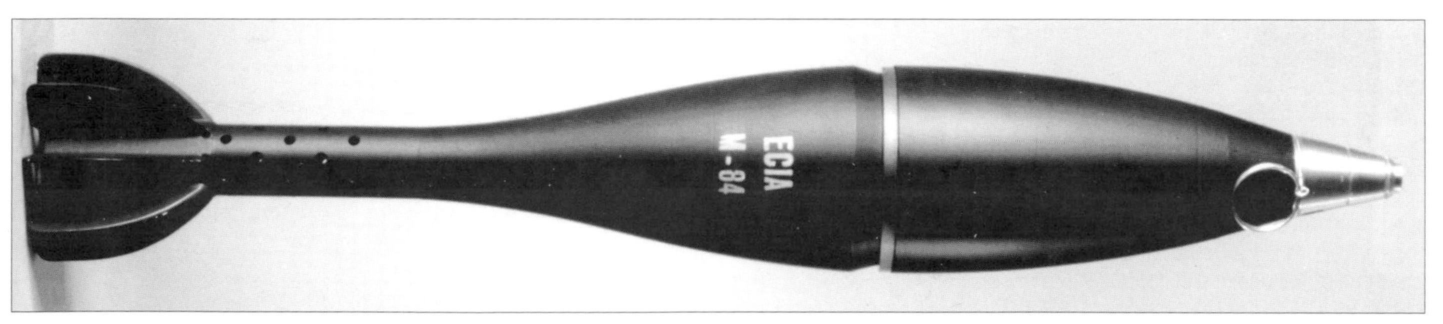

81mm Illuminating Bomb M68 France

Not all the bombs fired by mortars are high explosive; one of the most useful on today's battlefield is the illuminating bomb. Not only can it provide general illumination, but it can also be dropped behind distant tanks to silhouette them for the anti-tank missile operators.

The M68 bomb by Brandt Armaments of Paris is a good example of how these bombs work. It was adopted by the British Army for the 81mm L16 mortar.

In the cutaway photograph, it can be seen that the tail of the bomb is a cup-shaped forging into which the usual type of tail boom and fins are screwed. The rest of the body is of light sheet steel, secured to the tail by a few shear pins, and a time fuze is fitted into the head.

Inside the tail is another cup, with a compressed spring beneath it, and in this cup is a folded parachute. Above the cup, occupying most of the body, is a canister filled with magnesium flare compound. Between the canister and the fuze is a tiny gunpowder charge. The fuze is set to a time calculated to burst the bomb several hundred feet above the spot to be illuminated. When the fuze functions, it ignites the charge of gunpowder. This explodes; the flash from the explosion ignites the flare compound and the rise in pressure from the explosion causes the shear pins holding the body and base together to break. The compressed spring then pushes up the secondary cup and the parachute is ejected into the air. It opens and drags the burning canister from the fore body of the bomb and the parachute and light now begin to descend slowly. The pieces of the bomb fly off and land at random. The light lasts for about one minute, depending upon the height at which the bomb bursts, and provides illumination amounting to approximately 750,000 candle power.

DATA
Calibre 3.19in (81mm)
Length overall 16.42in (417mm)
Weight as fired 7.7lb (3.5kg)
Number of charges Primary + 5
Muzzle velocity 774ft/sec (236m/sec)
Range 3,500 yards (3,200 metres)

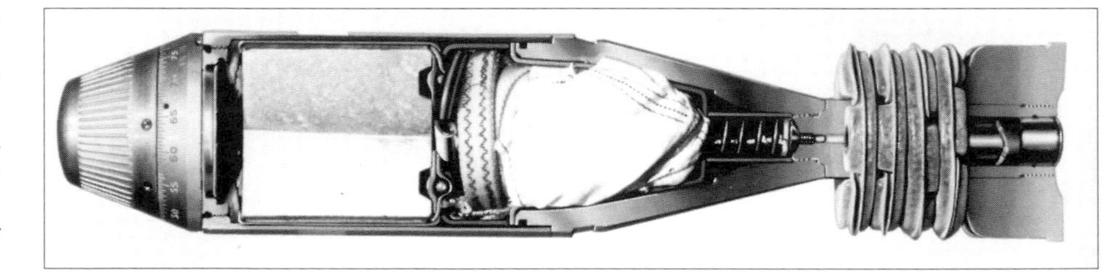

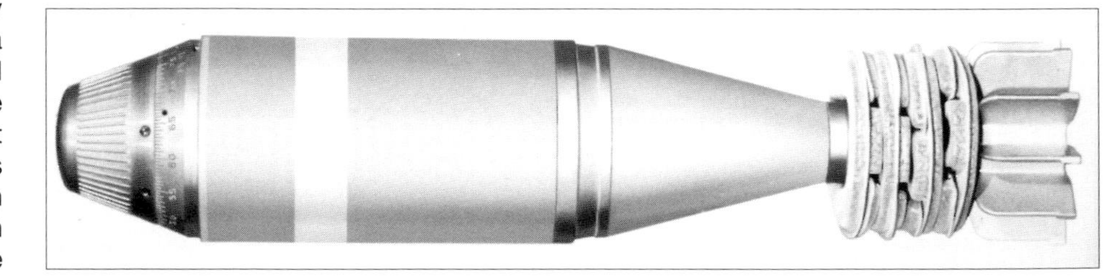

120mm Rocket Assisted HE Bomb M77 Serbia

This shows another approach to the rocket assisted method of obtaining greater range. The bomb is shaped like an artillery shell, with three machined bands around the body and with a squared off base. Below the base is a tail unit with four fins which are folded forward, like a jack knife; these fins are aerodynamically shaped and slightly canted so that when they extend in flight they induce a slow spin which aids stability of the bomb. Behind the tail unit is a cartridge holder, the usual perforated tube with a primary cartridge and six secondaries.

The body is divided into two compartments. The forward one, occupy-ing some two thirds of the body, is filled with TNT and carries a fuze in the nose. The rear third is filled with a rocket motor, venting through the base. This vent is closed by the cartridge container and, like the French design, has a selector unit inside.

Operation is similar to the French bomb. If rocket assistance is not required, then the bomb is drop loaded and fired. After leaving the barrel the fins fly out into the airstream and a small ejector charge blows off the cartridge container. If the rocket is selected by twisting the tail unit, then the charge also ignites a delay fuze which burns through and ignites the rocket on the final part of the bomb's upward flight. The rocket boosts the velocity and increases the maximum range from 5,300 metres to 9,400 metres.

DATA
Calibre 4.7in (120mm)
Length overall 30.0in (762mm)
Weight as fired 30.1lb (13.65kg)
Number of charges Primary + 6
Muzzle velocity 1,007ft/sec (307m/sec) with rocket
Range 10,280 yards (9,400 metres) with rocket

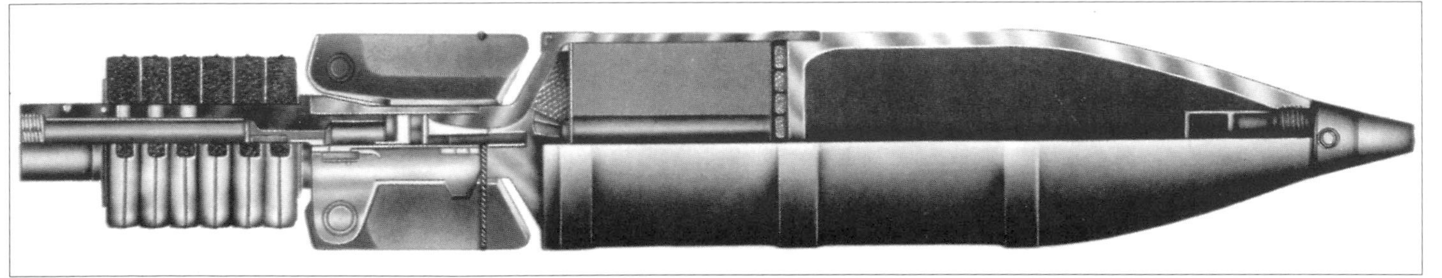

120mm Rocket Assisted HE Bomb PEPA France

This interesting bomb shows how the French increased the range of their standard smoothbore 120mm mortar from 4,250 metres to 6,550 metres.

The bomb has a steel body with a conventional impact fuze in the head. There are a number of gas check rings on the bourrelet and the body is also grooved, probably to assist in fragmentation. Inside, the body is filled with a Hexogen/TNT explosive and into this is inserted an insulated steel tube which carries inside it a stick of rocket propellant. The tube and the end of the bomb are closed by the tail unit, which is hollow, has a venturi jet for the rocket and behind this has a 'selector'. The tail unit has four fins which are folded for loading, but which fly out into the airstream after the bomb has left the mortar. Finally, loosely fitted into the rear of the tail unit, is the cartridge holder, with a primary cartridge inside it and the secondaries clipped around it.

The bomb is muzzle loaded and drop fired in the normal way, but before loading the decision has to be taken whether to employ the rocket or not. If not, no more needs be done; drop it in, fire and the maximum range is 4,250 metres. If the rocket is required, then the tail unit is twisted half a turn clockwise. This sets the selector and the bomb is then loaded and fired.

In either case the initial action is the same. The cartridges fire, the bomb leaves the mortar, the fins extend. The cartridges have also ignited a small expelling charge between the selector and the cartridge holder and shortly after the bomb leaves the muzzle this charge explodes and the cartridge holder is thrown off. If the rocket has been selected, this explosion also ignites a delay fuze which burns through and, shortly before the bomb reaches the highest point of its flight, the rocket ignites. As a result, the bomb continues upwards to a greater maximum height and thus the range increases.

DATA
Calibre 4.7in (120mm)
Length overall 29.85in (758mm)
Weight as fired 43.65lb (19.8kg)
Number of charges Primary + 7
Muzzle velocity 787ft/sec (240m/sec)
Range 7,163 yards (6,550 metres)

The PEPA rocket assisted bomb.
1) fuze;
2) explosive filling;
3) bomb body;
4) rocket chamber;
5) rocket propellant;
6) venturi;
7) optional delay;
8) selector lock;
9) spring operated tail fin;
10) tail assembly;
11) primary cartridge;
12) secondary charges.

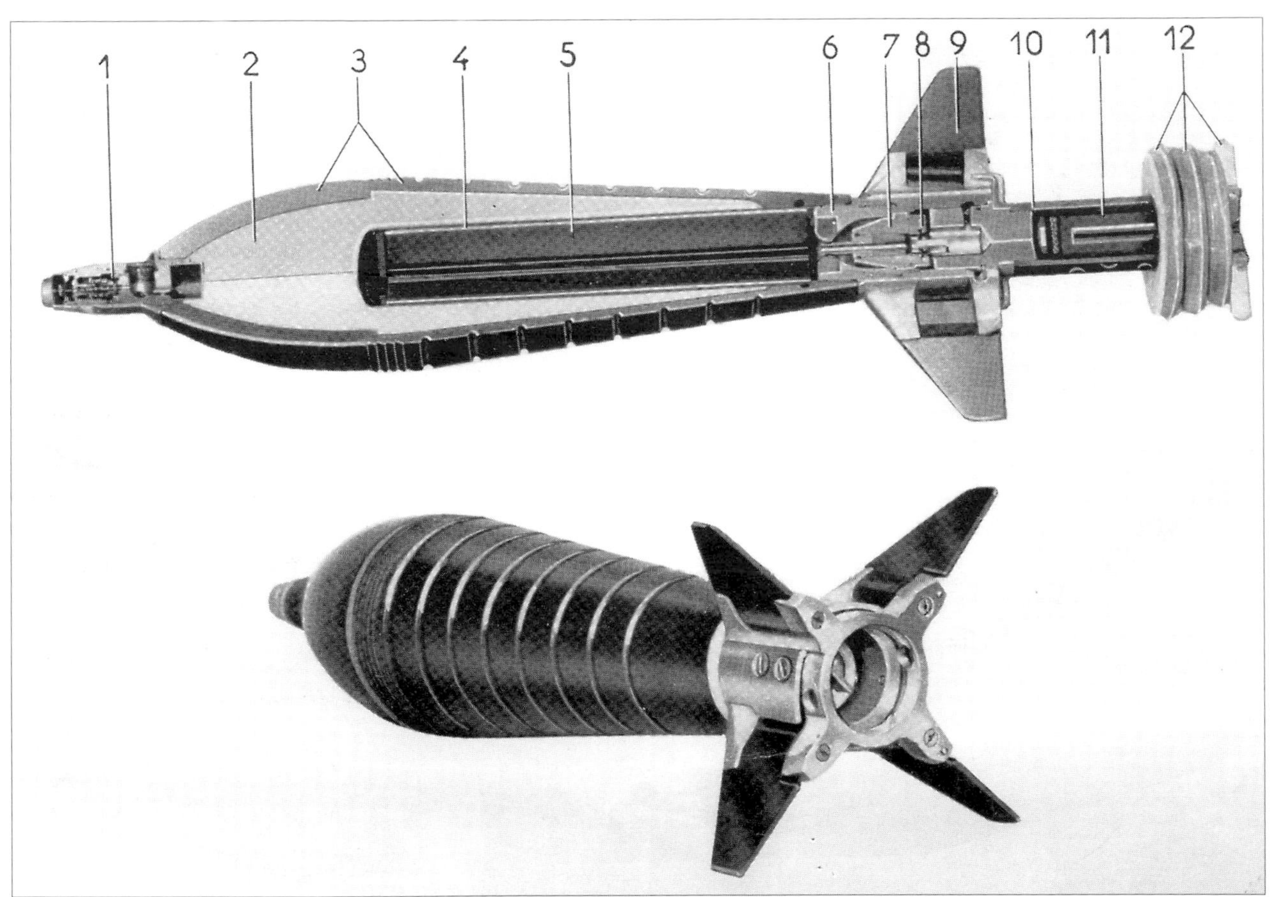
1
2
3
4
5
6
7
8
9
10
11
12

107mm HE Sub-Munition Bomb GRM20 — Greece

This was the first sub-munition bomb to be designed for a mortar and caused interest when it appeared in the late 1980s. It is designed for the American 4.2 inch (107mm) M2A1 or M30 rifled mortars, which are used in several countries.

As described in the introductory text, this bomb has a steel body shaped like an artillery shell, with a saucer-like copper cup under the flat base. Below the cup is a steel pressure plate and below that is the cartridge carrier, a simple perforated tube. A primary cartridge goes into the base of this tube and the secondaries are in the form of sheets of smokeless powder stitched together in bundles and clipped around the tube. On loading, the bomb slips down the rifled bore with no difficulty. On firing, the explosion of the charge forces the pressure plate up and squeezes the saucer-shaped copper plate outwards to dig into the rifling and spin the bomb.

Inside this bomb are 20 'Grenade M20G' bomblets; these are simply small steel cylinders containing a shaped charge, a parachute and a fuze. A time fuze in the nose of the bomb is set to burst the bomb open some 300 metres above the target area. At the set time the fuze ignites an expulsion charge which blows off the base of the bomb, allowing the grenades to fly out and spread their parachutes. The spin of the bomb causes the bomblets to spread out and fall to the ground to detonate, covering an area of about 7,000 square metres with lethal fragments. These bomblets will also penetrate up to 60mm of armour plate, should they strike a tank or other hard target.

An advantage of the rifled mortar is that ordinary spin actuated artillery fuzes can be used and this bomb allows the use of either mechanical or electronic time fuzes.

DATA

Calibre 4.2in (107mm)
Length overall 26.37in (670mm)
Weight as fired 28.44lb (12.9kg)
Number of charges Primary + 7
Muzzle velocity approx 984ft/sec (300m/sec)
Range 6,015 yards (5,500 metres)

(Right) The bomblet functioning against armour plate.

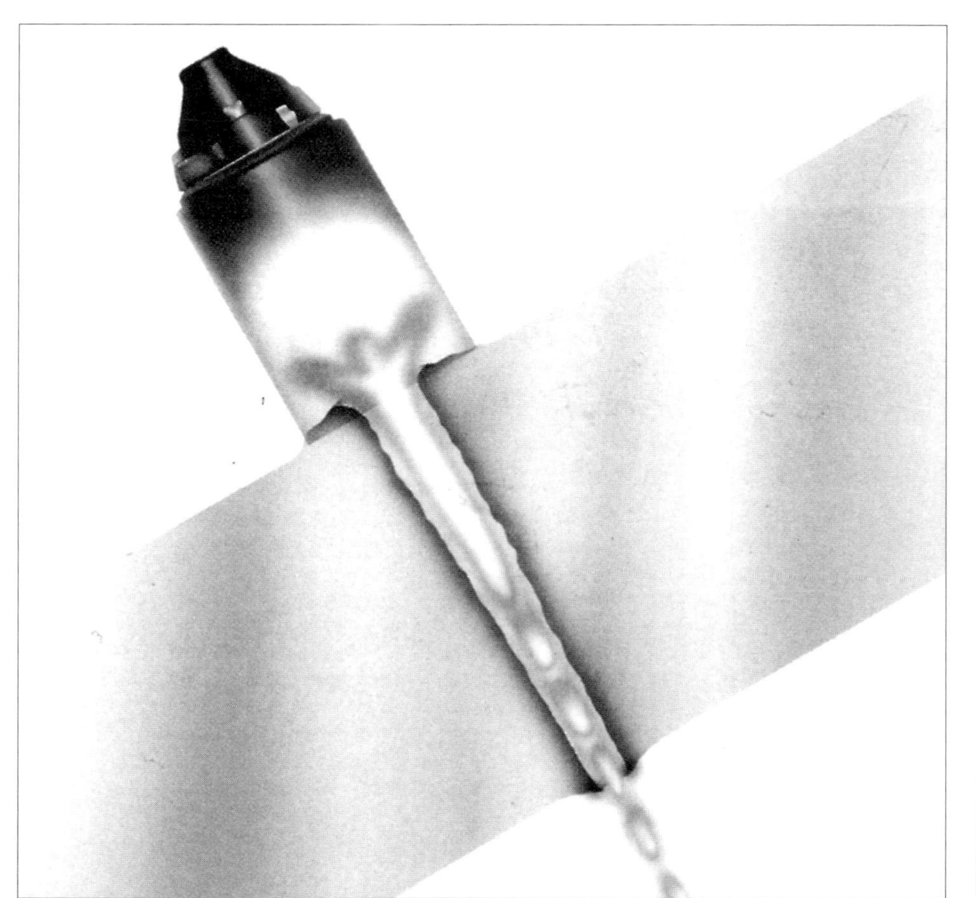

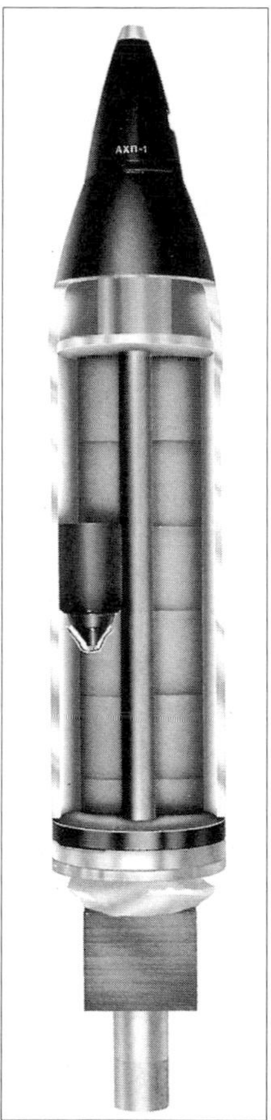
АХП-1

81mm APFSDS Shot SP81 France

An 81mm mortar is not the weapon one might expect to find firing high velocity armour piercing, fin stabilised, discarding sabot shot, but the Brandt company designed a 'gun mortar' for fitting into armoured vehicles which can function either as a conventional mortar, firing at angles of 45° elevation and above, or as a low trajectory gun. The weapon can be muzzle loaded or breech loaded, whichever may be the more convenient, though breech loading is more usual since the gunners are then protected by the armour. To fire as a mortar the breech is opened, the conventional mortar bomb loaded, the breech closed, the weapon elevated and the bomb fired. In the gun role however, the ammunition is entirely different.

This round resembles a round of tank ammunition, with a cartridge case in the mouth of which is fixed an APFSDS projectile. The cartridge is provided with spring clips to hold it in the breech in the correct position when the gun is at a low angle of elevation and the weapon is fired directly at the target. The supporting sabot is discarded at the muzzle and the sub-projectile, a hardened dart with fins, flies to the target. It is capable of defeating 90mm of armour at 1,000 metres range; not enough to deter a main battle tank, but quite sufficient to deal with an infantry fighting vehicle or an armoured personnel carrier.

DATA
Calibre 3.19in (81mm)
Length overall 20in (510mm)
Weight as fired n/a
Number of charges One, fixed
Muzzle velocity 2,953ft/sec (900m/sec)
Effective range 1,000 yards/metres

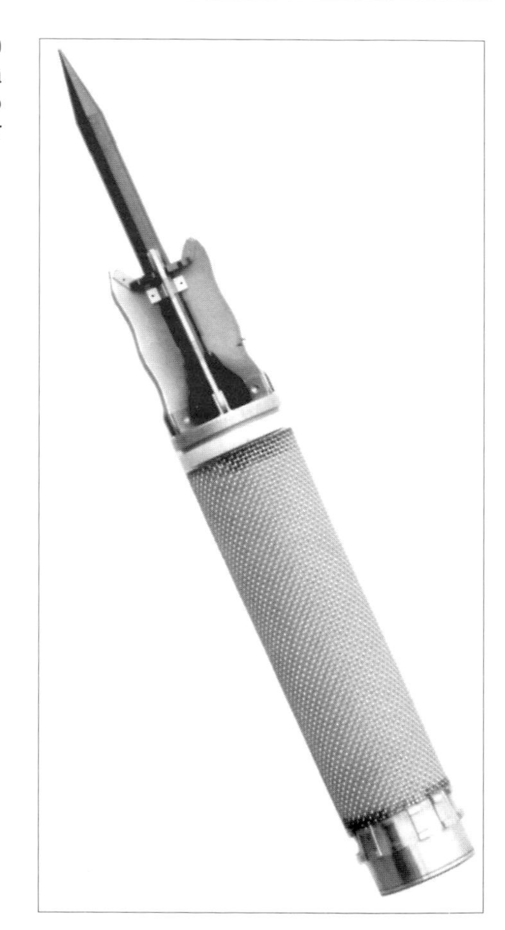

An 81mm gun-mortar mounted on a French AMX10P infantry combat vehicle.

120mm HE Improved Bomb HEI-L | Germany

This bomb was designed by the Rheinmetall company as a means of attacking hard targets such as tanks or field fortifications from above, generally the weakest part of the structure.

Externally, the bomb appears to be the usual type, with a streamlined body, light alloy tail unit and several gas check grooves around the body. Internally, however, it has some unusual features. The greater part of the body is taken up by the high explosive filling and the body is made of high grade steel so as to obtain thinner walls and thus the maximum capacity for explosive. The top of the charge does not reach to the nose, but is closed by a convex tungsten plate. There is a base detonating fuze in the bottom of the body and the nose carries a proximity sensor. Below the sensor and above the tungsten plate is a small 'separation charge'. The bomb body and the tungsten plate are grooved internally in order to promote their disruption into optimum sized lethal fragments.

The bomb is loaded and fired in the normal way. As the bomb approaches the target, at about 17 metres distance the proximity sensor ignites the separation charge. The charge blows off the sensor and nose cap so that the tungsten plate is exposed; this explosion also decelerates the bomb and this sudden check is sensed by the base fuze, causing it to detonate the main explosive charge. This then blows the tungsten plate forward, rupturing it into fragments as it does so and causing a cloud of fragments to be directed downwards at very high velocity to penetrate any hard target. At the same time the body of the bomb is fragmented and distributed around the bomb in the form of anti-personnel fragments.

DATA
Calibre 4.7in (120mm)
Length overall 23in (586mm)
Weight as fired 28.66lb (13.0kg)
Number of charges Primary + 9
Muzzle velocity n/a
Range 6,780 yards (6,200 metres)

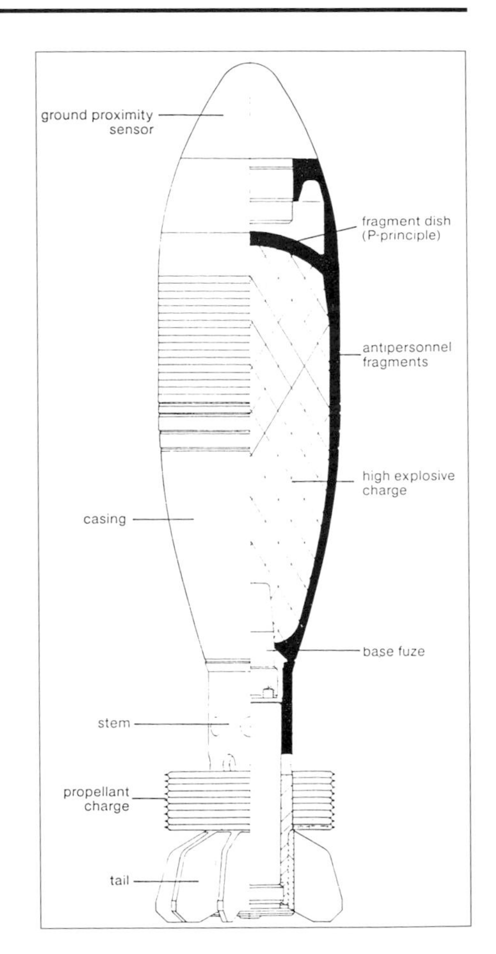

It is surprising what a well directed mortar bomb can do to a tank.

120mm STRIX Anti-Armour Guided Bomb — Sweden

Strix is a complete 'fire and forget' terminally guided missile which can be fired from any 120mm smoothbore mortar and is capable of destroying virtually any type of armoured vehicle. Using a mortar to attack tanks carries some advantages: in the first place the armour protection of the tank is always thinner on the top surfaces than it is on the sides and in the second place the near vertical fall of the bomb tends to avoid problems from smoke screens and other forms of concealment.

Strix was developed by Bofors and Saab Missiles from 1984 to 1990 and entered service with the Swedish Army in 1996. A long, cylindrical projectile with a rounded nose, it has four curved fins folded up against the body, which deploy after firing to stabilise the bomb in flight. Before loading it is fitted with a tail unit which carries the usual sort of primary and secondary propelling charges and it can also be fitted with a rocket booster to increase the maximum range.

Inside the body there is a powerful shaped charge at the rear end; running forward from this is a hollow tube, providing a space in which the shaped charge jet can develop and accelerate. Surrounding this tube are the various parts of the electronic guidance system and a ring of side facing thruster rocket motors. The nose of the bomb is filled with an infra-red seeker head.

After firing, the bomb leaves the mortar and the cartridge carrying tail unit drops clear, allowing the curved fins on the body to swing out into the airstream. If the rocket booster is fitted, this will ignite as the bomb approaches its vertex and boost the bomb to a higher altitude, so increasing the range. Once the rocket motor is burned out, the empty rocket units falls away from the bomb. During the downward flight of the bomb the electronics switch on and the infra-red seeker begins to scan the area in front of the nose. The seeker feeds information to signal processing circuits which can analyse the data and distinguish a casual fire or a burning vehicle from an operating vehicle. Once a target has been detected, the seeker locks on to it and the guidance system determines the amount and direction of correction to be applied to the bomb. Side thrusting rockets are then fired as necessary to steer the bomb on to the new trajectory and this corrective process continues until the bomb is aligned with its target.

As the bomb nears the target a proximity fuze fires the shaped charge and the jet passes down the hollow centre of the bomb, blasts its way through the minor obstacle of the infra-red sensor and penetrates the target. The manufacturers do not offer any figures for penetration, but a 120mm shaped charge should be capable of defeating any main battle tank in current service.

DATA
Calibre 4.7in (120mm)
Length overall 31.89in (810mm)
Weight as fired 37.47lb (17.0kg)
Number of charges Primary + 6
Muzzle velocity 1,050ft/sec (320m/sec)
Range 4,375yards (4,000m); 8,200yards (7,500m) with rocket booster

(Right) Testing and programming a Strix mortar missile before firing.

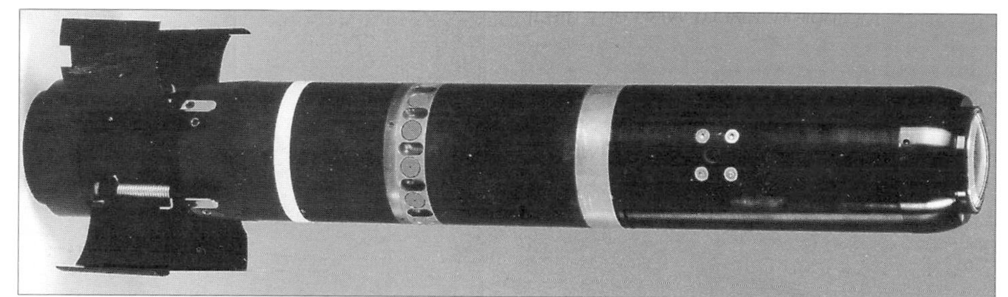

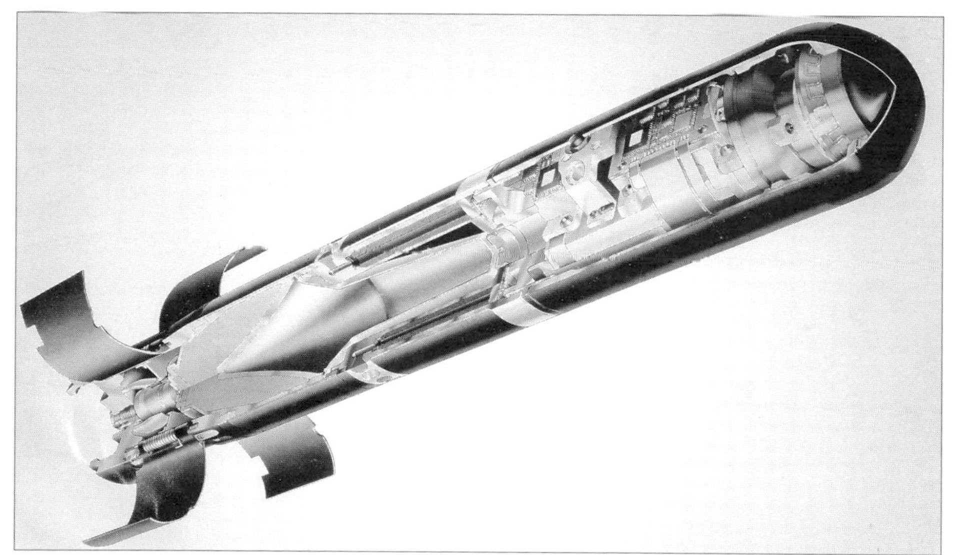

Anti-Tank Ammunition

There are only two satisfactory ways of attacking a tank: either by firing a very hard projectile at very high velocity in the hope of smashing a way through the armour plate, or discharging an explosive filled projectile at a lower velocity and relying upon the power of the explosion to defeat the armour protection. Neither way is as simple as it sounds, and not only because the designers of tanks have their own ideas on the subject.

The first tanks were protected by not much more than toughened boiler plate, which was sufficient to keep out ball bullets and grenade splinters, but later rolled steel armour became the standard material. This remained the standard material up to World War Two, with tank designers making the armour ever thicker. From the first arrival of the tank in 1916 it has been the subject of a see-saw battle: first the armour defeated the threat, then the threat defeated the armour, so the armour became thicker, so the threat increased... and it continues to this day, albeit in a rather more scientific manner.

The defeat of armour by artillery was something which had been fairly well mastered by the end of the 19th century, entirely due to the armoured battleship and, on the continent, the armoured fort. Two projectiles were in use by navies and coast defence guns, the armour-piercing (AP) shot and the AP shell. Both had hardened tips to pierce the armour; the shot was simply a solid piece of steel, while the shell had a small cavity in the rear end carrying a charge of high explosive and sealed by a base fuze. These enormous missiles – a 16 inch gun fired a shell weighing one ton – struck the armour with a force of several thousand foot tons and smashed their way through. The pieces of armour which broke off in the process were flung into the ship and the shot itself sometimes broke up to add to the missiles. The shell did the same thing but, if all went correctly, detonated just after it passed through the plate and dealt out death and destruction to the interior of the ship.

Surprisingly, this expertise was not called upon when the tank first appeared; it wasn't necessary. Armour-piercing rifle and machine gun bullets and ordinary field gun high explosive shells could defeat the early tanks without the need for special ammunition. As the tanks began to improve in the post-war world, the gun designers began to develop specialised weapons and, to go with them, the ammunition designers went back to the files on AP projectiles and began designing new ones. New in the sense of calibre and weapon, that is; they were simply fresh designs of AP shot and AP shell which were suitable for the new anti-tank guns, which, in the main, were of about 37–40mm calibre so that they could be easily man-handled by a few men, easily dug in and concealed.

After the outbreak of war in 1939 the contest between the armour and the gun intensified; guns became larger and more powerful and fired their shot and shell at greater velocities. This uncovered a defect which had never been seen before: when a steel shot struck an armoured target at velocities close to 3,000 feet per second, instead of penetrating, the projectile shattered to pieces due to the shock of the impact. The only solution was to use a harder material; the only harder material available was tungsten; and it was 1.6 times heavier than steel, so a shot made entirely of tungsten would never reach the necessary velocity to pierce armour.

We need not trace the various expedients tried along the way, but move straight to the final answer: the APDS or Armour Piercing Discarding Sabot Shot. In this the actual projectile that reaches the target is considerably smaller in diameter than the bore of the gun and is a solid slug of tungsten inside a streamlined sheath (it is easier to streamline the sheath than the core). The difference in diameters is made up from pieces of light alloy, called a 'sabot', which grip the core or 'sub-projectile' firmly so that the spin is transmitted from the rifling to the core, but this sabot breaks apart at the muzzle and flies off, leaving the core to travel by itself. The combination of core and sabot weighs a good deal less than a steel shot, so it reaches a higher velocity. By omitting the lightweight sabot and leaving only the very dense core, one is left with a projectile which has good flight characteristics, keeps its velocity well down range and penetrates a very thick piece of armour at the end of its flight.

There were good theoretical reasons to increase the weight of the core, but this ran into problems. There is an optimum spin rate for a given length of projectile and once the projectile becomes more than about eight calibres long, spinning doesn't do much to stabilise it. So, although it was an attractive idea to make the core longer and thus give it more weight and hence more impact at the target, it wasn't practical.

In the 1960s, the Russians came up with a new idea: make the core longer, give it fins and use a smoothbore gun. No rifling meant less drag and friction in the bore, so more velocity. The longer core meant more weight and better piercing ability and the fins provided the necessary stability. The only problem was that it wasn't particularly accurate at the first attempt and the Russians eventually had to put some slow rifling in the gun to assist stability. Their next attempt, a 125mm tank gun, showed considerable improvement and worked well with a smooth bore.

Once the Russians had shown the way, other designers got to work and the APFSDS (Armour Piercing, Fin Stabilised, Discarding Sabot) shot has now been brought into service in armies all over the world. The accuracy is all that can be desired and the penetrative power is formidable.

So much for the 'kinetic energy' method of defeating armour. The explosive energy method can be divided into two systems, HEAT and HESH. HEAT is High Explosive Anti-Tank and is the accepted shorthand for what is generally called the shaped charge or hollow charge. HESH is High Explosive Squash Head and is an entirely different approach. (HESH is also known, by the Americans, as HEP – High Explosive, Plastic, which gives a clue to its working.)

A HEAT shell is a thin-walled shell, the body of which is filled with high explosive and the nose left empty. The explosive is recessed into a cone at its front end and this cone is fitted with a copper liner. The fuzing system is arranged to detonate the explosive at the rear end, so that the detonation wave flows forward through the explosive and as it hits the copper liner, so it melts and deforms it, 'focussing' it into a thin jet of explosive gases and molten metal which is then projected forward at very high velocity in the region of 6,000 metres per second (20,000 ft/sec) to strike the armour. The combination of high temperature and high velocity simply melts away the armour and bores a hole through it.

One defect of HEAT is that the effect is highly localised; the jet is no more than about an inch in diameter. It is not unknown for older tanks, which had a certain amount of empty space inside them, to be struck by a HEAT projectile and suffer no ill effects; the jet simply went in through one side, passed across the empty space and went out on the other side. Modern tanks are less likely to get away with such a stroke of good fortune, since they are full of expensive electronics, ammunition, fuel and personnel. It would be a remarkable shot which failed to do some damage and the usually anticipated bonus is to aim the jet into the ammunition storage area and ignite the cartridges there. Naturally enough, the tank designers have their own views on this and go to great lengths to install fire quenching apparatus around the ammunition racks, but what might suffice against an ordinary fire will rarely move fast enough to extinguish a shaped charge jet conflagration.

The other defect of HEAT is that a shaped charge fired from a rifled weapon, and hence a spinning shaped charge, tends to dissipate some of its potential because the jet is flung outwards by centrifugal force, so the jet becomes thicker. This makes a larger hole, but one which is less deep, so against thick armour the jet can fail to penetrate, which is why fin stabilised HEAT shells fired from smoothbore guns are now on the inventory of most armies.

The HESH shell works in an entirely different way. The shell is thin walled and the body is filled with plastic explosive. There is a base fuze and the rounded nose is usually filled with some inert shock absorbing material. On striking the target the thin body collapses and allows the plastic explosive to fly forward and plaster itself against the armour plate like a sticky poultice. The base fuze then detonates it. This drives a high speed pressure wave through the plate, causing the inner face of the plate to fail; a 'scab' of metal is flung off at high speed, together with several hundred fragments of steel from the rupture, and these, whirling around inside the tank, plus the rebounding pressure wave, generally ruin the tank and its occupants beyond repair.

HESH is free from the problems which beset HEAT. Spin does not affect it and the effect on the target is far from localised. Given a big enough shell there is no tank which cannot be wrecked by HESH, but, generally speaking, for a given calibre of practical size, HEAT will defeat a greater thickness of armour than HESH.

Tank designers have gone to considerable lengths to attempt to protect their vehicles from the effects of these various type of attack. Simply piling on more armour is no solution, because the tank rapidly becomes so heavy that no engine will move it and few bridges will support it. Compound armour, which incorporates layers of steel, tungsten rods, plastic, ceramics, heat absorbing chemicals etc manages to stop some methods of attack. For example, planting tungsten rods inside armour steel will deflect an APFSDS piercing core so violently that it will generally snap, losing much of its mass and hence its piercing momentum. Plastics, ceramics and chemicals will absorb the high temperature HEAT jet and lessen its penetrative ability and will upset the shock wave set up by a HESH shell. The most recent counter measure is the application of 'Explosive Reactive Armour' (ERA) which is simply a steel box containing a quantity of high explosive and shaped so as to be fitted snugly around the turret and hull of the tank. The theory is that any attack of shot, HEAT or HESH, will detonate this charge of explosive and the counter blast will nullify the force of the attacking weapon. This system appears to work well on paper and in laboratory tests, but the limited amount of practical application which has so far been seen does not appear to bear out the theories and certainly the ammunition manufacturers do not appear to be deterred by ERA.

As with other sections, the entries which follow demonstrate a varied collection of anti-tank projectiles, from the basic to the most recent and innovative.

105mm OCC-105-F1 HEAT Shell

France

This is a remarkable projectile designed to overcome the problems associated with firing shaped charge shells from a rifled gun. This type of shell, which is uncommon, is known as the 'Gessner' projectile, from the name of the inventor.

It really consists of two shells. One is a shaped charge shell, the other is simply a hollow container with a driving band. The former fits into the latter and the two are separated by roller bearings. It can be seen from the drawing that the shaped charge shell is complete in itself, with a long cone shaped liner and a base detonating fuze. The fuze is connected to a piezoelectric unit in the nose, which is part of the inner shell assembly.

All this fits inside the outer body and there are roller bearings at the bottom, shoulder and nose to support the inner shell exactly on the axis of the outer shell. The outer shell also carries a tracer unit in the base for observing the trajectory.

On firing from a gun the outer shell takes up the spin from the gun's rifling, about 300 revolutions per second. But the amount of spin transferred to the inner shell is considerably reduced by the roller bearings allowing slip between the two and thus the inner shell has a very much reduced rate of spin, about 30 revolutions per second. On striking the target the piezoelectric unit is crushed and generates a pulse of electric current which flows down a wire and fires the detonator on the base fuze. The high explosive filling detonates, the wave runs forward and collapses the liner and the shaped charge jet forms and attacks the target. There is no particular obstacle in the forward fuze unit to obstruct the jet. Due to the reduced spin rate, there is very little dissipation of the jet due to centrifugal force and therefore good penetrative effect.

DATA

Calibre 105mm (4.13in)
Weight 24.14lb (10.95kg)
Length 18.3in (465mm)
Filling 1lb 11.5oz (780g) Hexolite
Muzzle velocity 3,280ft/sec (1,000m/sec)
Range <5,000 yards/metres
Penetration 14in (360mm)

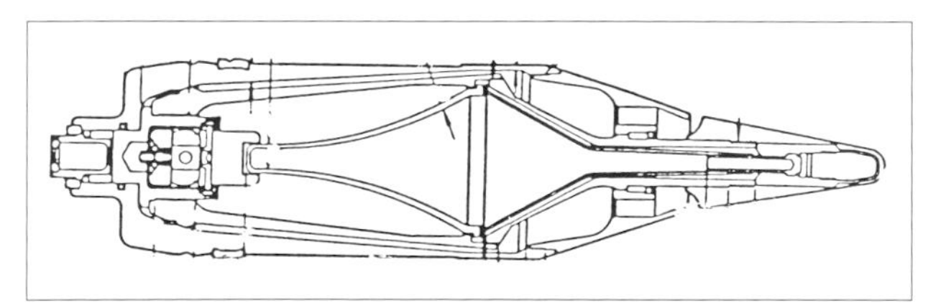

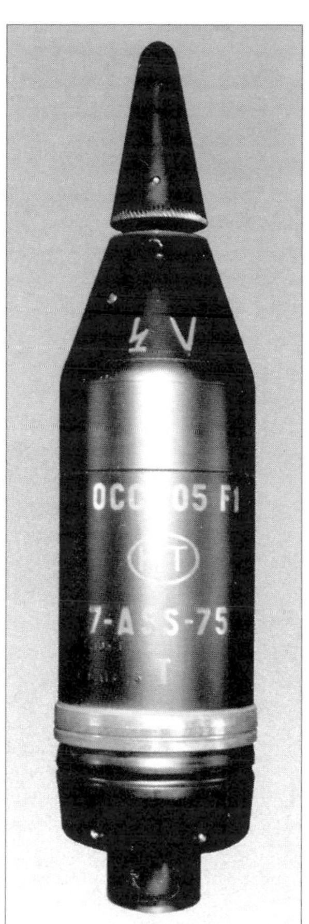

The Panzerfaust 60 — Germany

During World War Two the German Army relied upon anti-tank rifles at first, but soon found they were outclassed by improved tanks. In 1942 the US supplied the Soviet Army with the 2.36 inch 'Bazooka' rocket launcher; inevitably, some were lost to the Germans, who very rapidly copied it as their `Panzerschreck'. What the army wanted was something simple, which the individual soldier could carry, which didn't use much propellant and which could be thrown away after it had been used. This was the first time the concept of a 'disposable weapon' was put forward.

The answer was the 'Panzerfaust', developed by a Dr. Langweiler. It was a mild steel tube 31.5 inches long, with a simple sight and trigger mechanism on top. The bomb warhead was made of thin sheet metal, with a shaped charge and a hollow windshield in front. The tail of the bomb was a wooden rod, to which were attached four flexible sheet steel fins. These were wrapped around the tail rod and fitted inside the front end of the tube so that the large warhead was sticking out. Behind the tail rod, inside the tube, was a charge of gunpowder and, in the trigger assembly, a percussion cap.

To use the 'Panzerfaust', the first and perhaps the most testing requirement was to wait until the tank within 100 yards. Then the safety pin in the trigger block was removed, allowing the backsight to spring up; this had three holes in it, for 30, 60 and 80 metres range. The foresight was a pin on the edge of the bomb warhead. The firer

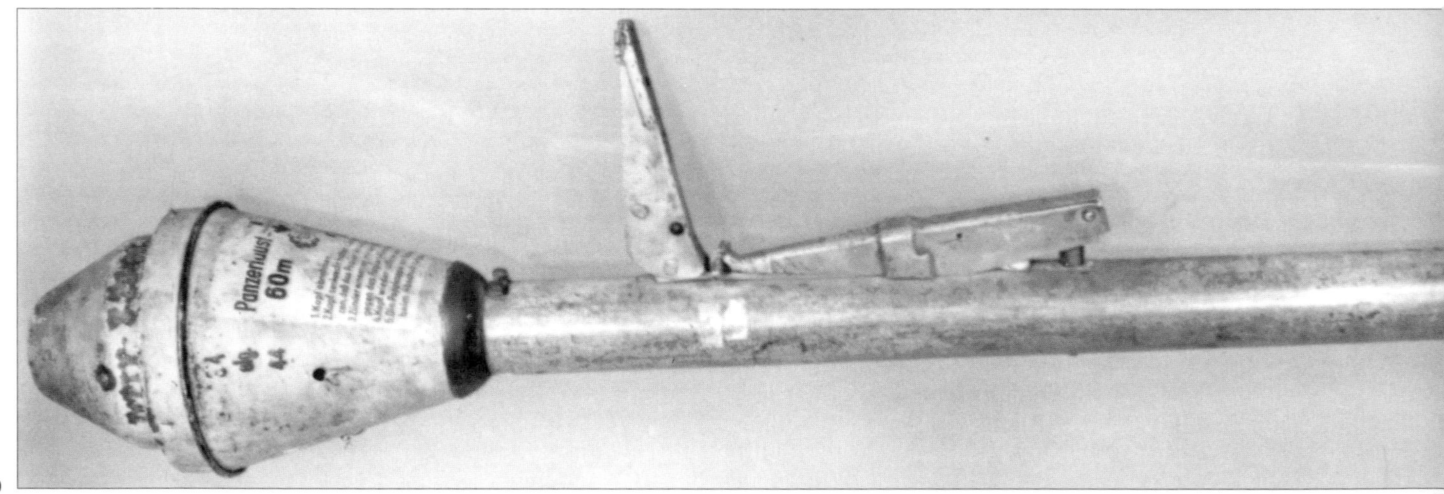

tucked the tube under his arm, placed his thumb on the trigger, aimed and pressed the trigger down. This drove the firing pin on to the percussion cap and fired the gunpowder charge. The blast blew the bomb out of the front end and also blew a charge of gas and smoke out of the rear end to give a recoilless effect. The bomb fins sprang out as it left the tube and when it struck the tank a base fuze detonated the explosive. The firer threw away the empty tube and reached for another.

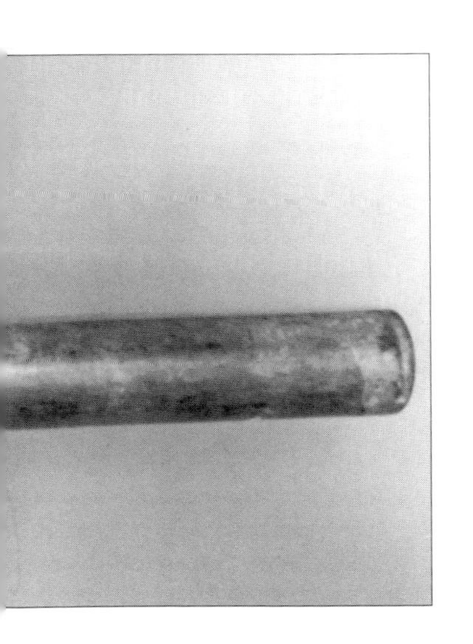

The first model was the Panzerfaust 30, with a range of 30 metres. This was rapidly improved to 60, then 100, 150 and finally 250 metres, though few of the latter were made before the war ended. It may sound primitive and it may look primitive, but it was lethal against any World War Two tank and it could still give a tank of the 1990s a severe shock.

DATA
Length of tube 31.5in (800mm)
Length of bomb 19.5in (495mm)
Weight, complete 7.5lb (3.5kg)
Diameter of bomb 5.9in (150mm)
Velocity of bomb 98ft/sec (30m/sec)
Penetration 7.8in (200mm)

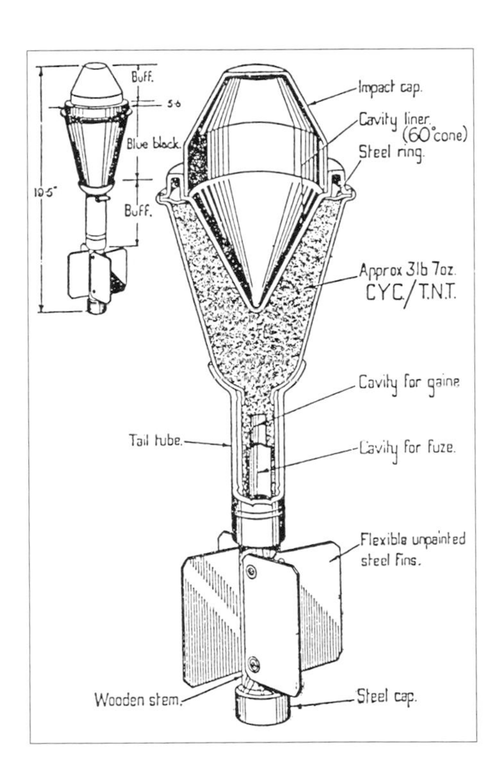

3.5 inch Rocket, HEAT, M28A2: The Super Bazooka USA

The original 'Bazooka' was a 2.36 inch calibre rocket, but by the end of 1944 the army was demanding something better and a 3.5 inch model was developed. The war ended before it could be brought into service and the design was shelved. It was brought back fairly rapidly when the Korean War showed that the 2.36 could make little impression on Russian T-34 tanks.

The launcher itself was no more than a light alloy tube 3.5 inches in diameter, with a pistol grip, a shoulder rest and an optical sight. For convenience in carrying, it came apart in the middle; when assembled it was 5 feet 1 inch long.

The rocket consisted of a warhead and a motor unit. The warhead, blunt nosed, contained the shaped charge, about 2lb of Composition B (RDX/TNT) with a copper cone in the front of it. The remainder of the nose was hollow and provided some stand off distance from the armour in which the explosive jet could form. Between the rear end of the warhead and the front end of the rocket motor was a base fuze, locked by a safety pin secured by a clip.

The motor was a thin steel tube which screwed into the base fuze at its front and carried a 3.5 inch ring fin assembly at its rear. At the rear end it narrowed into a venturi jet, through which the rocket gases emerged, and the motor, sticks of smokeless powder, had an electric igniter at the front end, connected to a contact ring around the fins.

To fire, the gunner put the launcher on his shoulder and took aim; his assistant took the rocket, removed the clip from the safety pin and pushed the rocket into the rear end of the launcher. There was a latch at the rear end of the tube which had to be lifted to permit the warhead to go in and the loader let it go as he pushed the fins into the tube. When he did so the latch dropped into a grooved contact ring around the rocket. Inside the tube the safety pin, which was spring loaded, was riding on the inside of the tube.

When the gunner pulled the trigger he actuated a magneto in the handgrip which sent current through the latch and the contact ring into the electric detonator in the rocket motor, completing the circuit through the safety pin and the launcher body. The rocket motor was burned out before the rocket left the end of the tube and as it came out so the safety pin jumped out and the fuze was armed.

The maximum recommended range for engaging a tank was about 200 yards; the sights were graduated up to 400 yards and with care and application it was possible to fire a rocket up to 1,200 or more yards against machine gun posts and similar obstacles. Although no longer used by NATO forces, there are still several armies with these weapons.

DATA
Calibre 3.5 inches (89mm)
Weight of launcher 12.1 lbs (5.5 kg)
Weight of rocket 8.9 lbs (4.04kg)
Length of rocket 23.55 in (598mm)
Warhead 31 oz (870g) Composition B
Velocity 312 ft/sec (95 m/sec)
Range Effective: 200 yards/metres;
maximum 1,200 yards
Penetration 7.8 inches (200mm)

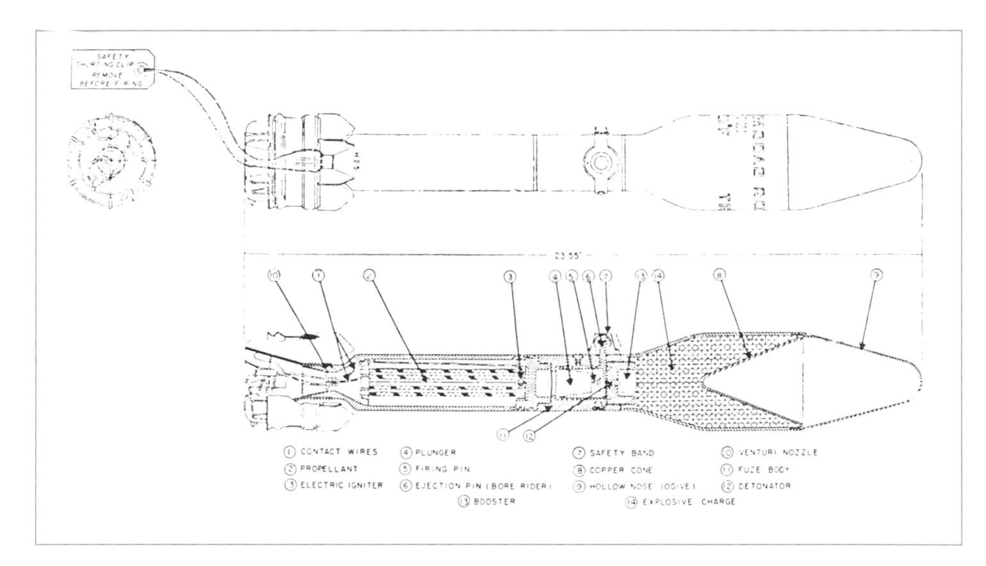

105mm APFSDS-T Shot C-437

This is a good example of a modern 105mm armour piercing fin stabilised discarding sabot shot for use in 105mm tank guns. It was developed by the Empresa Nacional Santa Barbara, the Spanish army arsenal.

The complete round consists of a brass cartridge case containing 5.85kg of smokeless powder, an electric primer in the base and the C-437 shot firmly fixed in the case mouth.

The shot consists of a tungsten alloy penetrator, the tip of which is covered by an aluminium windshield, inside which are a series of shock absorbing wads. The greater part of the penetrator is surrounded by a light alloy sabot which is split into three segments, lightly held together. Around the rear of the sabot is a plastic driving band which is free to rotate without transferring very much spin to the projectile. At the rear of the penetrator is a fin assembly with six alloy fins and, in the centre, a tracer element.

The round is loaded as a single unit and fired by an electric current directed into the primer via an insulated firing contact in the gun breech block. The entire sabot projectile is launched down the bore; the forward section of the sabot rides on the surface of the bore, while the plastic driving band digs into the rifling to seal the propellant gas behind the sabot and spin the whole projectile. At the muzzle the air pressure on the recessed forward section of the sabot causes the three segments to split apart and peel off sideways, falling to the ground some 400 yards ahead of the gun. The sub-projectile is now left to carry on to the target. On striking, the wads in the nose absorb some of the initial shock and act as a lubricant to assist the tungsten penetrator to pierce the armour. The penetrator makes a hole some 6 to 7cm (2.25 to 2.75 inches) in diameter and can defeat the standard NATO single heavy target (a 120mm armour plate set at an angle of 60° to the line of fire backed by 10 additional 10mm plates set 10mm apart) at a range of 5,000 metres.

DATA

Complete round weight 39.7lb (18kg)
Complete round length 36.5in (928mm)
Projectile weight 12.45lb (5.65kg)
Projectile length 17.1in (435mm)
Muzzle velocity 4,920ft/sec (1,500m/sec)

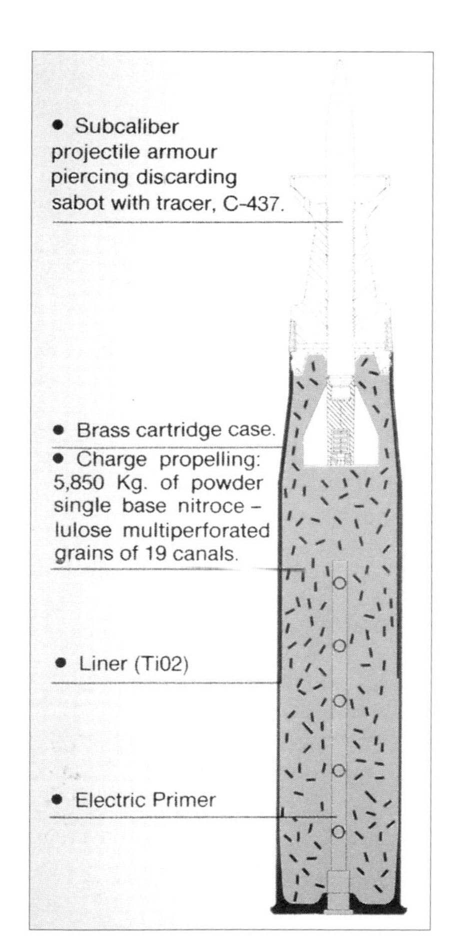
• Subcaliber projectile armour piercing discarding sabot with tracer, C-437.
• Brass cartridge case.
• Charge propelling: 5,850 Kg. of powder single base nitroce - lulose multiperforated grains of 19 canals.
• Liner (Ti02)
• Electric Primer

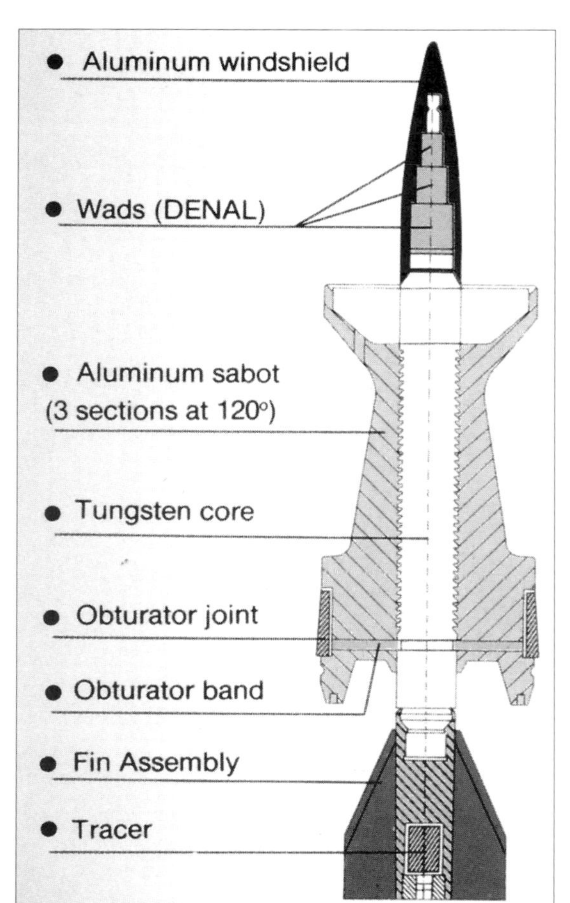
• Aluminum windshield
• Wads (DENAL)
• Aluminum sabot (3 sections at 120º)
• Tungsten core
• Obturator joint
• Obturator band
• Fin Assembly
• Tracer

MECAR 105mm TPDS-T Practice Shot M724 — Belgium

The best ammunition is of no avail if the man firing it doesn't know his job, and frequent practice is therefore necessary. Ammunition for tank guns, particularly sabot ammunition, is expensive; the heavy metal penetrator is costly in terms of both the material and the man hours involved in shaping it. There is therefore a need for a cheaper round of ammunition which will perform the same as the service round for range and accuracy. It doesn't matter that the round will not penetrate armour; it is sufficient that it can punch a hole in a wooden target screen.

The round shown here is a Target Practice Discarding Sabot, Tracer (TPDS-T) round for 105mm tank guns, made by the MECAR company of Belgium for supply to NATO forces, most of whom use the British 105mm gun in some of their tanks. It is based upon an originally British design and is a spun projectile which uses a `pot' sabot; the construction can be seen in the drawing.

The sabot consists of a cylindrical 'pot' in which the sub-projectile sits, located at its nose by a three piece belt lightly held together by a plastic band. There is a plastic driving band on the pot, concealed inside the cartridge case. The sub-projectile consists of a soft steel sheath carrying a tracer at the rear, inside which is a core of heavy steel with an aluminium nose pad. This weighs somewhat less than the tungsten penetrator usually used, but provided an adjustment is made to the sight, the trajectory resembles that of the service APDS shot.

On firing the whole assembly goes up the bore and spins; on clearing the muzzle centrifugal force throws off the three front segments, breaking the plastic retaining band, and air drag on the pot causes it to slow up and fall from the rear of the sub-projectile.

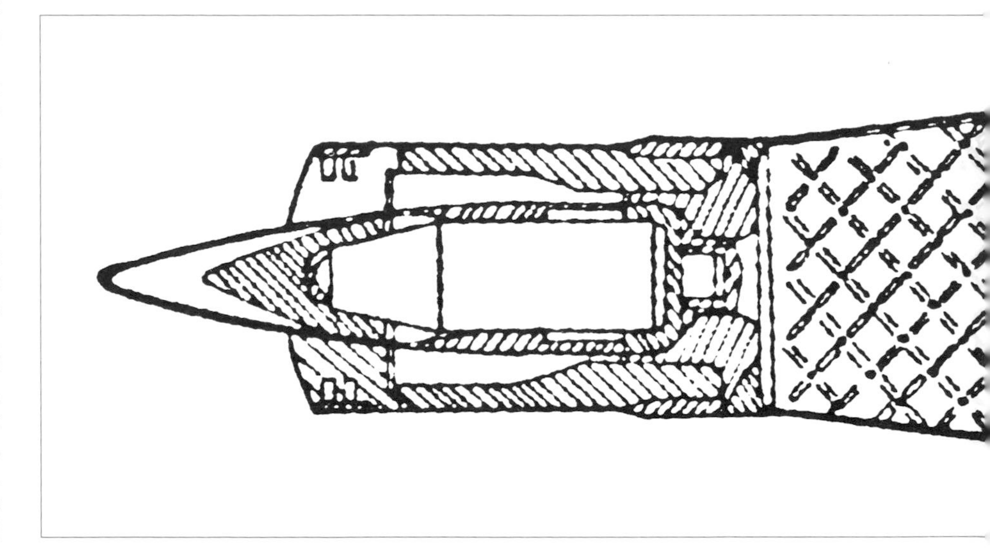

DATA
Complete round weight 29.5lb (13.4kg)
Projectile weight 6.8lb (3.1kg)
Muzzle velocity 5,035ft/sec (1,535m/sec)
Range In excess of 7,650 yards (7,000 metres)

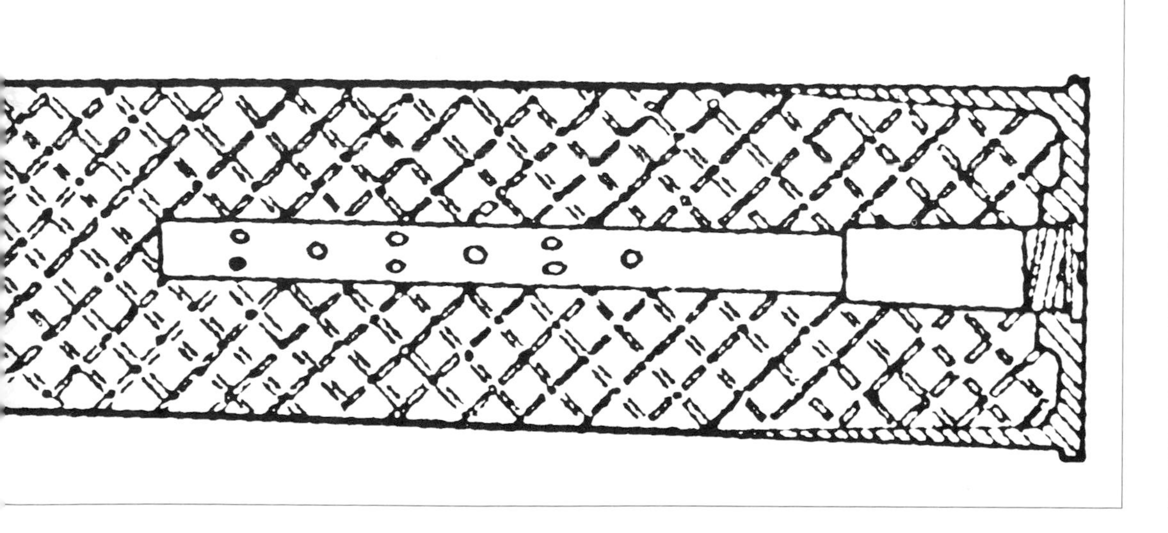

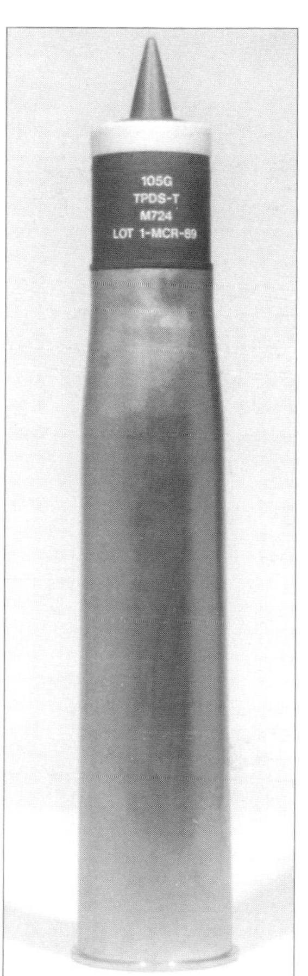

105mm TPFSDS-T Practice Shot DM-148 — Germany

If training is a problem with APDS rounds, training with APFSDS ammunition is a much bigger problem. The streamlined shape of the APFSDS sub-projectile and the exceptionally high velocity mean that if the gunner misses his target, or if the sub-projectile ricochets off the ground and into the air, it can travel for anything from 30 to 50km before it comes down again. No existing anti-tank gunnery range has sufficient space to catch such a loose round.

This problem faced the German Army in the 1970s and they passed the problem to the Diehl company, experienced ammunition specialists. Their solution is this training round DM-140 which outwardly resembles the standard APFSDS DM-23 service round, but has some important internal differences.

As the sectioned picture of the projectile shows, it is to the same general construction as the service round, a long and thin fin stabilised dart held in a petal sabot which is discarded at the gun muzzle. But the sub-projectile is not a single piece of tungsten. Instead, it is of steel, in two pieces. The front piece contains four pyrotechnic delay charges, while the rear piece carries the fins and has a tracer and a single pyrotechnic delay unit. The four charges in the forward section are provided with set back detonators, and the rear unit may have a set back detonator or it may be lit from the tracer element.

The round is loaded and fired in the same way as the service round, but as it starts up the bore the sudden acceleration causes the set back detonators to fire the pyrotechnic delay units. Up to about 3,000 metres range the sub-projectile behaves exactly as would the service round and can be used for target practice. At a point somewhere between 3,100 and 4,500 metres from the gun the pyrotechnic elements burn through their delay and fire a separation charge between the two halves of the sub-projectile. This immediately breaks in two and since the two elements are now unstable, they fall to the ground fairly quickly and well within a radius of 7,500 metres of the gun. The five pyrotechnic units are there simply as insurance; any one of them will fire the separating charge on its own, but having five ensures certainty of action.

DATA
Projectile weight 9.5lb (4.3kg)
Projectile length 16.8in (427mm)
Muzzle velocity 3,445ft/sec (1,050m/sec)
Operating range 3,400–4,920 yards (3,100–4,500m)

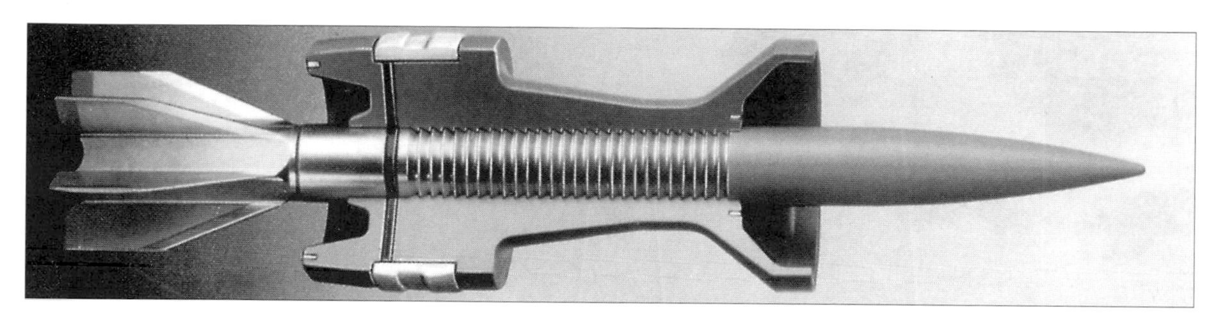

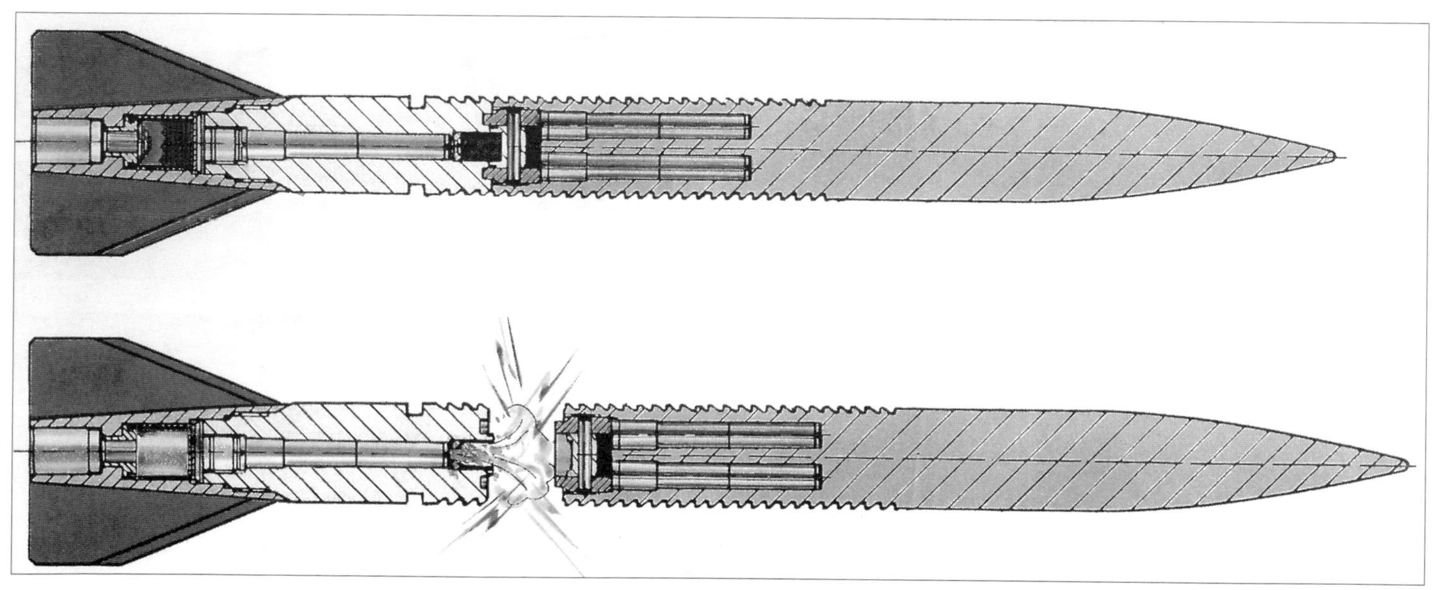

RAW: Rifleman's Assault Weapon

USA

This remarkable device, which is neither a grenade, nor a shell, nor a mortar bomb nor anything else which can normally be classified, is offered as a means of satisfactorily dealing with hard targets with nothing more than a rifle.

RAW was developed by the Brunswick Corporation of the USA as a result of requests for a powerful but portable weapon for fighting in built up urban areas where an enemy can protect himself by strong walls and sandbagged positions and the attackers can rarely get tanks or artillery to bear on them. A heavy charge of explosive, accurately placed, is the only solution and RAW provides it.

The projectile is a 140mm diameter sphere, to which is attached a rocket nozzle. Inside the sphere is the warhead, its fuze and a rocket motor. It is spin stabilised, using its rocket for propulsion and to generate spin, and the thrust is such that it flies on a flat trajectory for about 200 yards before assuming a ballistic curve. The launcher is an attachment to the rifle, a metal bracket which slips over the muzzle and does not interfere with the normal use of the weapon. The soldier inserts the rocket nozzle into the firing tube on the attachment and fires an ordinary ball cartridge. Some of the propellant gas passes down a tube in the attachment to drive down a firing pin in the rocket motor and ignite it. Some of the first rocket blast passes through two side jets on the launch tube and gives the rocket its spin and at the same time the thrust launches the sphere from the bracket.

A number of alternative warheads have been developed for RAW. The standard warhead is a squash head (HESH) filling of 2.2lb of high explosive, actuated by an impact fuze; this will blow a 14inch diameter hole in an eight inch thick reinforced concrete wall. The RAT (RAW Anti-Tank) warhead is a shaped charge with an optical proximity fuze which will detonate the charge at the correct distance from the armour and will defeat up to 18 inches of armour plate. The 'Flying Claymore' warhead carries thousands of tungsten pellets and blasts them out at high velocity in a most destructive swath; this warhead can be fired to over 2,000 yards range. Other warheads which have been tested and proved include incendiary and smoke types.

The RAW has been taken into use by the US Marine Corps and is being further developed as part of the US Army's MPIM (Multi Purpose Individual Munition) programme.

DATA

Weight of projectile and launcher 8.5lb (3.86kg)
Payload of projectile warhead 2.51lb (1.14kg)
Length of projectile 12in (305mm)
Diameter of projectile 5.5in (140mm)
Effective range 7.8 in (200m)
Maximum range 2,190 yards (2,000m)
Maximum velocity 567ft/sec (173m/sec)

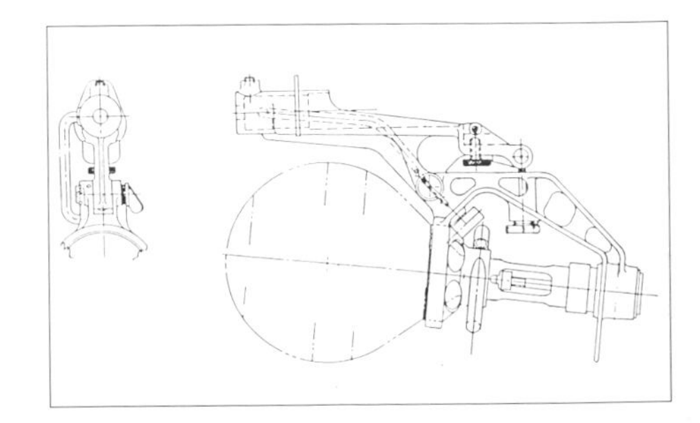

(Left) The RAW rocket in flight.

(Below) The result of firing RAW against reinforced concrete.

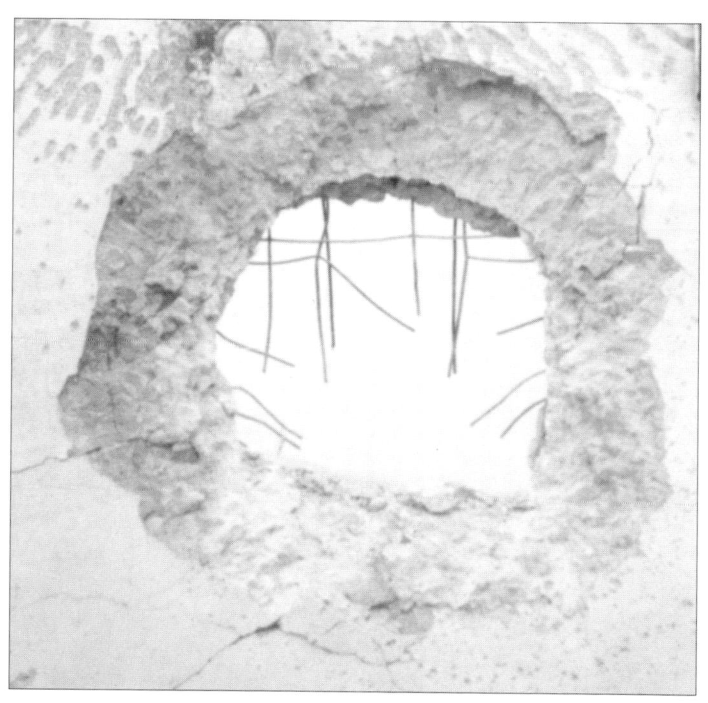

105mm HEAT-MP-T M830 — USA

The M830 is a fin stabilised HEAT shell for firing from the smooth bored 120mm tank gun; it also serves to introduce the combustible cased charge.

As tank guns increased in size, the cartridge cases also got bigger and they eventually became something of a menace in the confined space of a tank turret as they were ejected, smoking hot, from the gun breech. They had to be removed, otherwise there would soon be no room for the crew and the prospect of opening the hatch in the middle of a battle to throw out the empty cartridge cases was not enticing. Automated systems for catching the empty cases and ejecting them through special ports generally failed sooner or later and caused even more trouble.

The answer was to make the actual cartridge case out of some material which would burn and so act as a useful addition to the propelling charge. There is a short metal 'stub case' at the base, to provide the necessary gas seal in the gun breech, but the remainder of the case is actually made from a mixture of nitrocellulose, paper and resins and is similar to heavy cardboard in appearance. It is however, a low explosive and combines with the normal propellant inside the case to generate propulsive gas. It is therefore entirely consumed in firing and the only thing to come out of the breech after firing is the stub case, which is no more than a few inches long.

The projectile is cylindrical, with a long nose probe which carries the initiator for the base fuze and provides the desired stand off distance. Inside the shell body is the usual sort of shaped charge and behind it is a long tail boom with a fin assembly and a tracer at the end of it. The long tail boom is necessary in order to position the fins well behind the body so as to be in the air stream and thus have the desired stabilising effect. The fins also have wide shoes on their outer edge which ride on the gun barrel and, together with the body of the shell, ensure correct centering and stability of the shell as it goes up the bore.

DATA
Projectile length 33.1in (842mm)
Projectile weight 29.8lb (13.5kg)
Complete round length 38.6in (981mm)
Complete round weight 53.4lb (24.2kg)
Muzzle velocity 3,740ft/sec (1,140m/sec)
Effective range 2,500 yards/metres

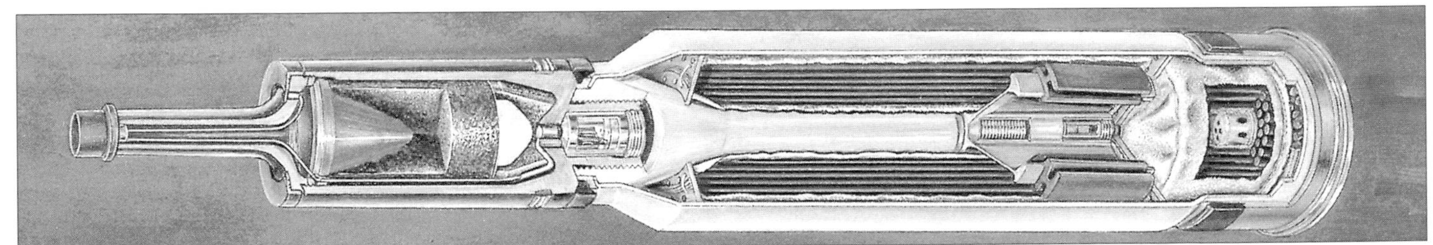

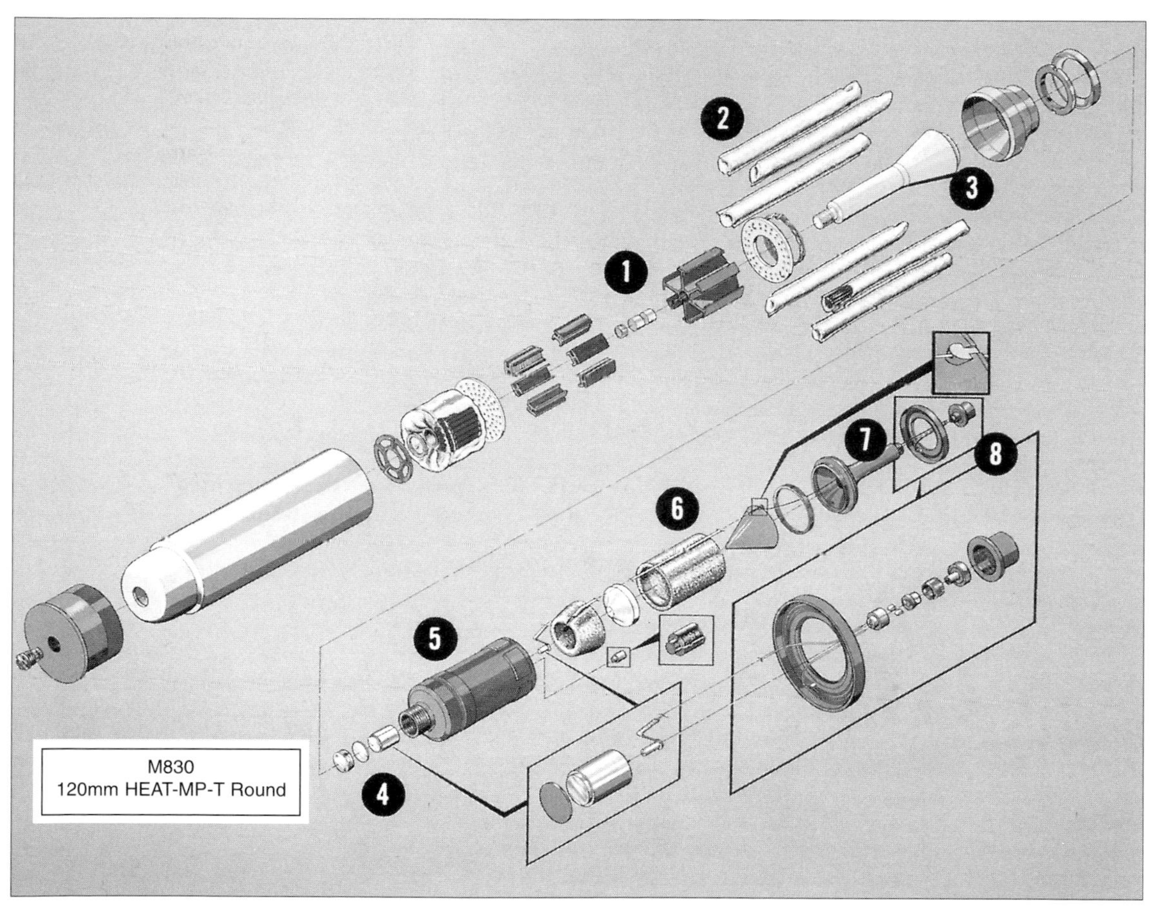

The M830 round:
1) tail and tracer;
2) stick propellant;
3) tail boom;
4) base fuze;
5) shell body;
6) shaped charge;
7) nose spike;
8) impact sensor and switch.

105mm HESH Cartridge DM-512

Germany

This is a NATO standardised round which is based upon the American HEP-T M393; a similar round is also made in France as the NR 132, in Belgium as the M393A2 and in Greece as the M467. It can be fired from any standardised 105mm tank gun.

The cartridge is the conventional brass cased type, filled with granular propellant powder and fitted with an electric primer.

The HESH shell consists of a thin metallic body with a blunt nose. The nose portion is filled with an inert shock absorbing material, the remainder of the shell with a plastic high explosive. There is a cavity in the rear end of the filling into which a base detonating fuze fits. The fuze itself is screwed into the base cap of the shell, which, in turn, is screwed into the shell body. There are two copper driving bands on the outside of the body and a tracer element is screwed into the rear end of the base cap.

On firing the tracer ignites and burns for about four seconds of flight, supposing the target to be so far away. On impact the shock absorbing material prevents a shock induced detonation of the explosive and the thin casing splits open to allow the plastic filling to be thrown forward to stick tightly on to the surface of the armour. The base fuze, initiated by the sudden deceleration, now detonates the explosive and a violent shock wave is driven into the plate. This reaches the inner surface of the plate and rebounds; this increases the stress in what is known as the 'pressure bar effect' and the result is to cause a complete failure of the metal about one inch from the inner surface. An area roughly the same size and shape as the original pad of plastic explosive is released by this failure of the metal structure and is flung off, at something in the order of 1,000ft/sec velocity, into the interior of the tank. In addition, the area of failure disintegrates into thousands of small fragments which are also flung in at high velocity. The net result is total destruction as the heavy scab of metal bounces around inside of the tank, accompanied by a cloud of fragments. There is also a secondary, external effect, as the explosion of the plastic HE shatters the shell body and base and distributes them around the outside of the tank.

DATA
Complete round weight: 46lb (20.9kg)
Complete round length: 37in (940mm)
Projectile weight 24.9lb (11.3kg)
Projectile filling 5.04lb (2.286kg) plastic HE
Muzzle velocity 2,417ft/sec (737m/sec)
Maximum range 10,390 yards (9,500m)

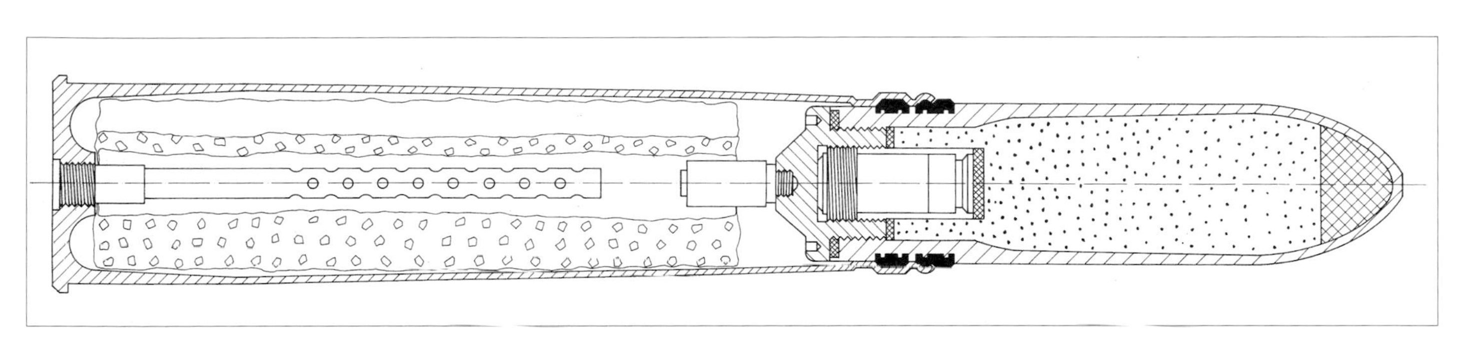

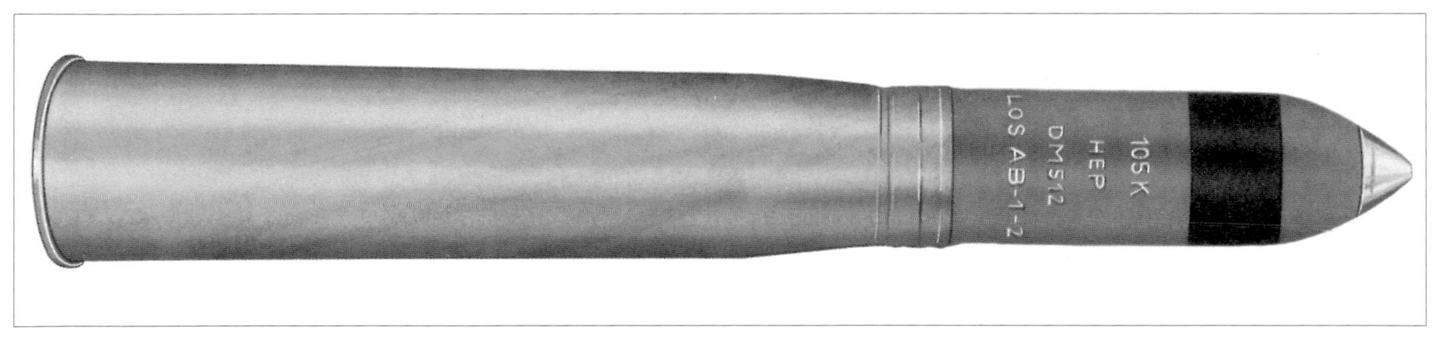
105 K
HEP
DM512
LOS AB-1-2

Artillery Ammunition

The design of artillery ammunition showed very few signs of progress for almost 20 years after World War Two; such new designs as did appear were merely variations on previous designs and were little different in substance from ammunition which had appeared in the wake of World War One. But in the 1960s the rush towards missiles was beginning to slow down and the potential power of nuclear weapons virtually guaranteed that they could never be used. So firstly the military began rethinking their artillery doctrines and secondly the various high technology manufacturing companies, with now under-occupied design staffs, decided to look elsewhere in the military armoury to see where they could introduce new ideas.

The basic fundamentals of artillery ammunition have not changed: burning a low explosive in a confined space behind a projectile and launching it from a gun which is rifled so as to give a stabilising spin to the projectile. The projectile will follow a calculable ballistic trajectory due to air drag and gravity and it will require a fuze to produce the desired effect at the target. Having carefully defined all that, we will shortly see that there is not one of those fundamentals which has not been altered in some design of modern ammunition and sometimes more than one in the same design. When aerial bombs began taking on technology, somebody defined them as being either 'iron bombs' or 'smart bombs'. In the same way we might refer to 'iron artillery' and 'smart artillery' and the consolation is that there is still plenty of iron artillery about, so that our fundamentals of ammunition are still fairly safe.

One of the principal western military concerns during the Cold War was the Soviet tank threat, with perhaps three or four Soviet tanks for every western one. The Soviet artillery threat was even greater, though very little publicity was ever given to it. As a result, the earliest applications of technology were directed towards countering these two threats by the use of special artillery projectiles; firstly to stop tanks well away from the line of contact and secondly to deal with artillery at long ranges before it could develop into a tangible threat. As these projectiles went through their development stages, they had a useful side effect: they directed the attention of the technologists to the guns that were to fire these new projectiles and that led to a number of new ideas in gun design. When new and novel ideas occur in gun design, the next thing to happen is that these ideas render new artillery tactics possible. New tactics often suggest new targets and so a new ammunition idea comes round and the circle is completed. The advances in artillery equipments and techniques since 1970 have been greater than anything that had happened in the previous 70 years.

Broadly speaking, artillery ammunition falls into one of two groups: that for guns which use metal cartridge cases and that for guns which do not. Ammunition for guns which use cartridge cases can also be subdivided. If the shell (or shot) is fixed in the mouth of the case and the 'round' is loaded in one unit, then it is 'fixed ammunition'. If the shell and case come as separate items and the shell is loaded and rammed first and the cartridge case inserted afterwards, then it is 'separate ammunition'. The Americans, for their 105mm howitzer, prefer a method in which the case and shell are supplied as one unit but loosely fitted together, so that the shell can be taken out and the propelling charge changed, after which the shell is put back into the case mouth and both are loaded as one unit. This is classed as 'semi-fixed ammunition'.

The cartridge case performs a number of functions. It contains the charge of propellant powder, carries the primer which ignites the charge, protects the contents from rain and damage, supports the shell in the case of fixed and semi-fixed rounds and, most importantly, it expands slightly in the chamber as the charge explodes and thus seals the breech of the gun against the propelling gases escaping to the rear. (This sealing action is properly called 'obturation'.) Cartridge brass, an alloy of 70% copper and 30% zinc, is traditional in this role because it has the desired quality of being able to expand to perform the obturation task and then, once the internal pressure drops, contracts so that the empty case can be extracted. It is also easy to draw and form and it is resistant to corrosion from damp. But copper and zinc are always in great demand in wartime and as a result all the combatants in World War Two, sooner or later, began looking at substitute materials, with the result that today cartridge cases are frequently made from steel.

Guns which don't use cartridge cases have their charge inside a cloth bag; this means that the cartridge no longer performs the obturation task and this has to be achieved by a resilient pad forming part of the breech block. It also means that the primer has to be a separate item and fired in a miniature breech system called the 'firing lock' which delivers a flame through the central vent in the breech block to strike the bagged charge. The bag was traditionally a coarse silk cloth, but by 1939 this was becoming both scarce and expensive, certainly in the quantities demanded by artillery, and substitute fabrics had to be found. The main requirement was that the fabric should be completely consumed in the explosion of the charge and not leave smouldering residue in the chamber on to which the incoming charge might carelessly be loaded. After that it needed to be moderately waterproof and not prone to be consumed by tropical insects. Fortunately the late 1930s saw the beginning of man-made artificial fabrics and some of these proved adaptable to this role.

Smokeless powder is difficult to ignite and so it is necessary to provide some assistance for the flash which enters the chamber from the primer. This is achieved by stitching an 'igniter' to the end of the cartridge bag, a simple round cloth bag filled with gunpowder. It is usually coloured red, so that the loader can see which end of the bag had to go next to the breech face so that the flame from the primer will hit the igniter. This takes fire and ignites the contents of the bag in a satisfactory manner.

Smokeless powder is usually referred to as 'cordite', but cordite has not been used for many years; it was a mixture of nitro-cellulose and nitro-glycerine which was very powerful, but exceedingly hot and tended to wear away the interior of the gun. This has been replaced by cooler burning mixtures which also generate less flash and smoke.

Fuzes are classified in a number of ways; they can be nose or base fuzes, according to where they are inserted into the shell. Nose fuzes are more common, since they are easily adjusted and they are the first part of the shell to strike the target and thus initiate the action of the shell. Where it is necessary to break through a hard target before bursting the shell, or where it is essential to initiate the shell filling from the rear end, then base fuzes are used.

Fuzes can also be classified as impact, time or proximity. Impact can be sub-divided into direct action, where the fuze has to strike the target, or graze action where the deceleration of the shell as, for example, if the shoulder 'grazes' the ground, is enough to start the action. Time fuzes were traditionally designed to burn a regulated train of pyrotechnic powder; these were gradually replaced by mechanical fuzes in which some sort of clockwork mechanism did the timing more accurately. These are now being completely superseded by electronic fuzes which rely on a simple micro chip or a resistor condenser circuit to provide the necessary timing. Proximity fuzes sense the presence of the target and detonate the shell at the optimum point to produce the desired effect. They are usually controlled by a radio circuit which transmits a weak signal and listens for it to return from the target; once the return is of sufficient strength, the shell is detonated. But this type of fuze can also operate by acoustic or optical sensing methods, although these are uncommon in artillery applications.

Proximity fuzes were first developed for anti-aircraft gunnery, where a near miss was more likely than a direct hit, but they are now used for bursting shells over the top of troops in trenches and also for detonating shaped charge and similar anti-armour projectiles at the optimum distance from the armour to allow the charge to have its best effect. It is perhaps worth mentioning that jamming a proximity fuze is not quite as easy as is often thought; modern proximity fuzes usually have a timing circuit which does not switch on the transmitter until it is almost at its target, so that if a jammer is used and does burst the shell successfully, it will burst it over the enemy's head anyway.

With the essentials established, let us now see how they are being put together, starting with some `iron artillery'.

105mm HE M1

USA

This must be the most produced and most widely distributed artillery round in history; it was developed in the early 1930s for the American M1 field howitzer, has stayed in use through several upgraded designs of gun, has had guns designed around it in other countries and has been manufactured in over a dozen countries. It is also the only current example of a semi-fixed round of ammunition. The shell and cartridge are supplied together, but the shell can be pulled free from the cartridge to permit the propelling charge to be adjusted. This charge is in seven bags of smokeless powder, tied together with string and the string can be snapped and bags removed so as to produce the desired range and trajectory. The use of adjustable charges allows high velocity flat trajectory shooting or low velocity high trajectory shooting so as to drop the shell down behind cover. It also avoids using a bigger charge than is necessary and thus saves wear on the gun.

The shell is a simple and basic high explosive shell, forged from 23 ton steel and filled with Composition B, which is a mixture of RDX (also called Hexogen) and TNT. This gives a detonation which will break the body of the shell up into small high speed fragments, and cover a fairly large area around the point of burst, and it will also develop sufficient blast to severely damage any structure it strikes. The fuze issued as standard is an impact and delay fuze, originally the M48/M51 series, now the M557 and similar types. These all use a direct action striker and detonator in the nose, with an inertia pellet and detonator in the body of the fuze. There is also a selector switch on the side of the fuze which, when turned to 'Delay' will stop the flash from the impact detonator having any effect and thus allows the inertia pellet to fly forward as the shell suddenly decelerates and fire the delay detonator. This slight delay allows the shell to penetrate light cover, such as the roof of a house, and burst inside. In addition, a number of different time or proximity fuzes can be used with this round.

DATA
Calibre 105mm
Length of round 31.1in (790mm)
Weight of round 39.9lb (18.1kg)
Weight of shell 33lb (14.97kg)
Payload 5.07lb (2.3kg) Composition B
Propelling charge 2.73lb (1.24kg) in 7 zones
Muzzle velocity 1,548ft/sec (472m/sec) (M101 howitzer)
Maximum range 12,325 yards (11,270m) (M101 howitzer)
Lethal area 340 sq.yards (285 sq.metres)

105mm M1 howitzer round.
1) fuze; 2) shell body; 3) booster charge; 4) explosive filling; 5) cartridge case; 6) propelling charge bag; 7) primer.

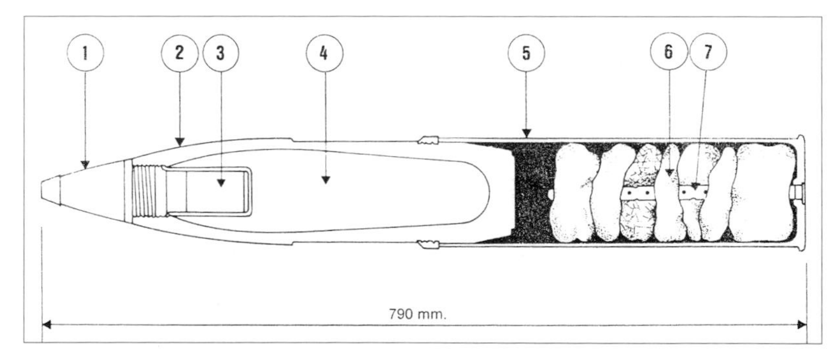

105mm BE Smoke M84 USA

This introduces the concept of the 'carrier shell', a shell which itself has no target effect but which carries something inside it which will perform the necessary task at the target. In this case the task is to lay a smoke screen.

For many years after its introduction in World War One the smoke shell was more or less the same as the HE shell, but filled with white phosphorus (WP) and a small charge of explosive sufficient to break the shell open at the target. The WP began generating smoke as soon as it met the air, the only drawback was that the smoke was hot and heated the air around it, which rose and took the smoke with it, so that instead of a low clinging screen, a towering pillar was produced and thus a lot of smoke shells were needed to provide a decent screen. In the late 1930s the British worked out a new method and this American shell is based upon the British design.

The shell is simply a steel tube with a driving band and a time fuze. Below the fuze is a small charge of gunpowder and a loose steel plate with a hole in the centre. Beneath this and filling up the shell body, are three steel canisters filled with a cool burning smoke mixture. The canisters have a hole through the centre and they are held in place by a baseplate either lightly screwed or pushed and pinned into the base of the shell body.

The fuze is set to burst the shell at the desired spot. It is fired in the normal way and at the set time the fuze ignites the gunpowder expelling charge. This explodes and the flash passes through the hole in the steel plate and down through the centre of the canisters, igniting the smoke mixture. The explosion pressure pushes downwards on the plate and on the canisters and shears the screw or pins holding the baseplate, ejecting the canisters into the air. They fall to the ground and emit smoke for up to 90 seconds, smoke which clings to the ground and forms a dense screen.

It is necessary to fire a number of these shells at a fairly rapid rate to build up the screen, after which it has to be 'stoked' by firing more shells at about one minute intervals until the screen is no longer required. Similar shells with the canisters filled with coloured smoke or coloured flare mixtures are used for indicating targets to aircraft or to observers on the ground or for other signalling purposes.

DATA
Calibre 105mm
Length of round 31.1in (790mm)
Weight of round 41.95lb (19.03kg)
Weight of shell 30.86lb (14.00kg)
Payload 3 smoke canisters containing 12.1lb (5.5kg) of mixture
Propelling charge 2.73lb (1.24kg)
Muzzle velocity 1,548ft/sec (472m/sec) (M101 howitzer)
Maximum range 12,325 yards (11,270m) (M101 howitzer)

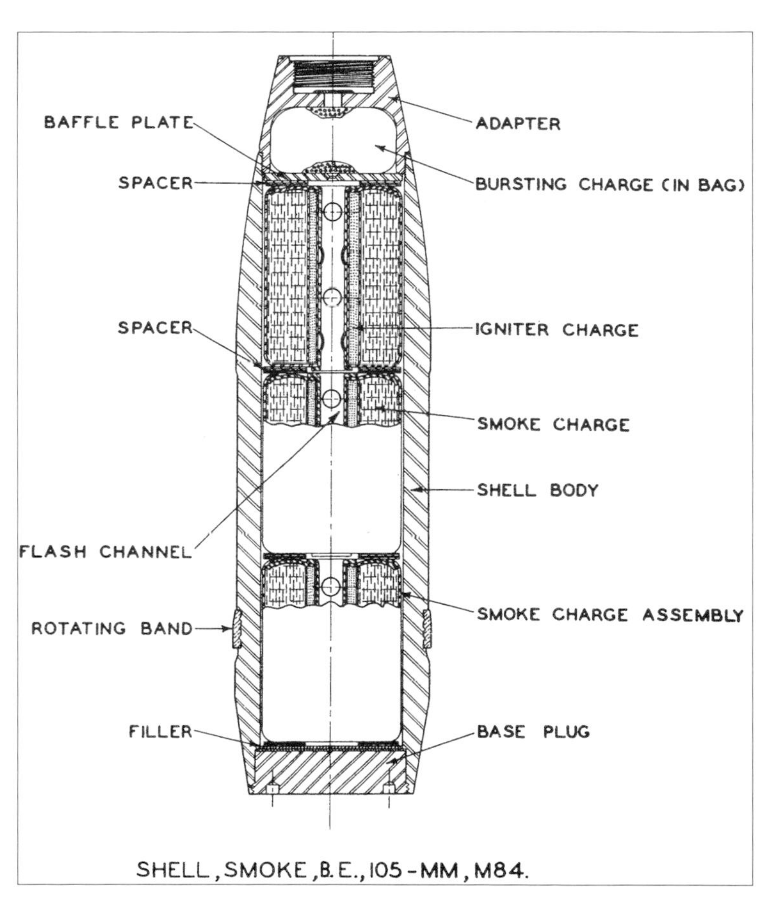

SHELL, SMOKE, B.E., 105-MM, M84.

105mm HE RA M913

USA

One of the first and most obvious ways to obtain greater range from a conventional gun or howitzer was to add a rocket booster to the shell. This was pioneered by Germany during World War Two as an off-shoot of their widespread rocket development programme, but although they put a few designs into service the problems were not really solved until the 1960s. The principal defects are firstly that putting a rocket inside a shell reduces the explosive payload, so that the target effect is reduced, and, secondly, if the shell happens to be slightly unstable and yawing off course at the instant the rocket fires, then it will be driven off its predicted trajectory and miss its target.

The round shown here, for the American 105mm M119 howitzer, solves the first problem by adding the rocket unit to the basic shell instead of building it in and cures the second drawback by careful attention to the flight characteristics.

The cartridge differs from that of the standard HE round in having a non-adjustable propelling charge. The shell has a relatively thin walled body with a greater explosive payload than normal and attached to the rear end of the shell body is the rocket assembly. This is loaded with a composite solid fuel, a delay unit and a `rocket separator cap' and it also carries the driving band for the complete projectile.

Before firing, the shell is removed from the cartridge case and the rocket separator cap is turned to either select rocket action or prevent it. If the selector is turned off, then the shell performs in the conventional manner and the rocket does not ignite. If the cap is turned on, then as the cartridge explodes, so the delay unit in the rocket assembly is ignited. This burns for 16 seconds of flight and then ignites the rocket motor to accelerate the shell and increase the range.

The accuracy of the shell when fired with the rocket assistance is claimed to be 0.35 percent of range and 1 mil for direction. This means that at 15,000 metres range the shell should land within a rectangle 15 metres wide and 53 metres long and with the high proportion of payload in the shell, this is a very satisfactory result.

A similar round, the M927, which differs in having the usual seven zone adjustable propelling charge, is provided for the M101 howitzer.

DATA
Range without rocket 15,750 yards (14,400 metres)
Rang with rocket 21,325 yards (19,500 metres)
Lethal area 613 sq yards (513 sq metres)

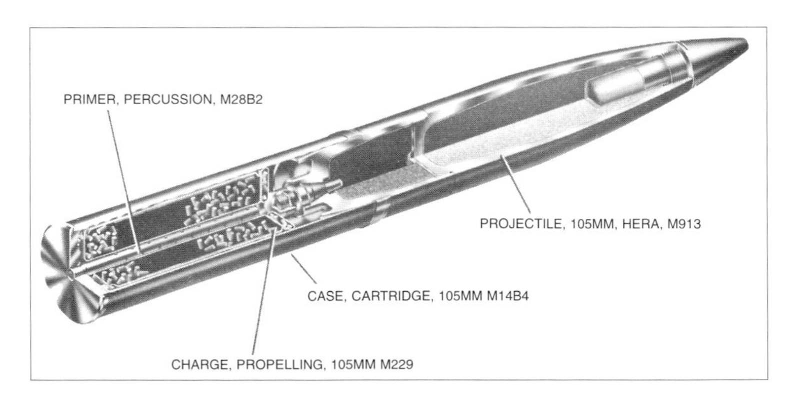

Extended Range Full Bore (ERFB)

This method of improving the range of shells can trace its origins back to experimental work carried out in Germany in 1942–45, when Dr. Banck of Rheinmetall Borsig was seeking the perfect projectile. His idea was to produce the best possible flight shape and then fit it with various supports and sabots to suit the bore of the gun. One of his methods was to attach stub wings to the tapering front section of the shell, wings which rode inside the bore and kept the shell steady and then acted aerodynamically to give the shell a little extra lift during flight. His experiments were not completed before the war ended and although reports on his work and specimens of his ammunition were available after the war, very few people appear to have taken much notice.

In the 1960s Dr. Gerry Bull, a Canadian researcher, began studying the problem of increased range and rediscovered Banck's idea of the stub wing. By this time the 155mm howitzer was the preferred artillery general support weapon and he therefore designed an entirely new 155mm shell incorporating stub wings, a hemispherical base and a long tapering body. Due to the long taper, the stubs were necessary to keep the shell aligned with the axis of the barrel as it was fired and they also provided lift. The hemispherical base reduced the amount of air drag behind the shell and also shifted the centre of gravity forward to make the shell more stable, and the tapering body gave it a very low drag factor. The net result was to increase the range of the US 155mm Howitzer M109 by no less than 38 percent, from 14.6km to 19.3km. With an entirely new howitzer having a longer barrel and designed to extract the utmost from the new shell, the gain was even more spectacular: from 17.8km firing a conventional shell to 30km firing the new shell, a gain of 68 percent.

The title of ERFB – Extended Range, Full Bore was chosen to indicate that the system did not depend upon a sub-calibre projectile and sabots, as had been tried several times previously, and the system has been adopted by several countries.

DATA (Typical)
Calibre 155mm (6.1in)
Weight 100.3lb (45.5kg)
Length 33.18in (843mm)
Payload 19.18lb (8.7kg) RDX/TNT
Muzzle velocity 2,943ft/sec (897 m/sec)
Maximum range 33,027 yards (30,200 m) (from 45 calibre barrel)

Extended Range, Full Bore, Base Bleed (ERFB-BB) South Africa

Towards the end of World War Two some British ordnance engineers fitted a tracer to a 9.2 inch coast artillery shell. When it was fired, the shell went a good deal farther than expected. The war ended before more investigation could be conducted and that particular trial was closed down. What had been achieved, in fact, was to provide the shell with a source of high pressure gas which filled the void behind it and thus reduced the base drag. As the air flows over the shell it re-unites some distance behind it and the gap is a vacuum which acts as a brake. By filling this with the gas from a burning tracer the drag was reduced and the shell went faster and further.

Nothing much resulted from this idea for several years, but in the 1960s it re-appeared as a method being suggested by Dr. Gerry Bull to give another boost to his ERFB shell. Beyond suggesting it and making a few experiments, Bull did little, but the South African company Armscor set about serious research and eventually perfected a workable system which has since been refined and copied by other manufacturers.

In many ways the system is similar to rocket assistance, except that the burning substance is designed purely to generate gas rather than thrust. The hemispherical tail end of the ERFB shell is removed and the base bleed unit is screwed on in its place, a task which can, if necessary, be done in the field. The unit is a steel container loaded with a block of inhibited propellant powder. It is ignited by the flash of the propelling cartridge and burns for about 30 seconds of flight. The gases produced pass through a central hole in the base at subsonic speed, fill the vacuum behind the shell and prevent the disturbed air from flowing inwards and creating turbulence which would otherwise slow the shell's flight. The amount of drag is thus reduced to between 50 and 75 percent of that found in conventional shells and increases the range by up to 25 percent without any significant change in accuracy.

ERFB and ERFB-BB shell design has been extended from high explosive shell to cover smoke, illuminating and other types of carrier shell, so that the advantages can be gained in every tactical situation.

DATA (Typical)
Calibre 155mm (6.1in)
Weight 105.15lb (47.7kg)
Length 33.9in (861mm)
Weight of BB unit 10.58lb (4.8kg)
Length of BB unit 4.7 in (120mm)
Payload 19.18lb (8.7kg) RDX/TNT
Muzzle velocity 2,936ft/sec (895 m/sec)
Maximum range 42,650 yards (39,000m) (from 45 calibre barrel)

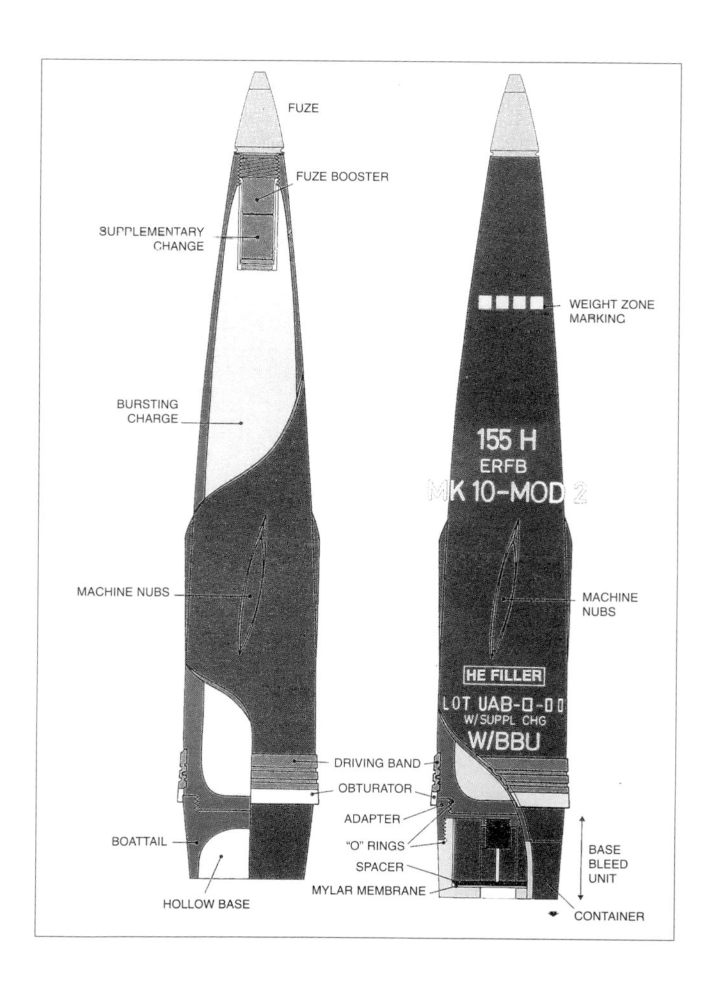
FUZE
FUZE BOOSTER
SUPPLEMENTARY CHANGE
BURSTING CHARGE
MACHINE NUBS
BOATTAIL
HOLLOW BASE
WEIGHT ZONE MARKING
155 H
ERFB
MK 10-MOD 2
MACHINE NUBS
HE FILLER
LOT UAB-0-00
W/SUPPL CHG
W/BBU
DRIVING BAND
OBTURATOR
ADAPTER
"O" RINGS
SPACER
MYLAR MEMBRANE
BASE BLEED UNIT
CONTAINER

Dual Purpose Improved Conventional Munition M864 USA

The base bleed system just described can be used with otherwise conventional projectiles and is not confined to the stub wing ERFB shell. This example, the DPICM, uses base bleed with a conventional shell shape, though the payload is far from conventional.

The M864 is an advanced model of one of the earliest designs of 'Improved Conventional Munition', the advancement being the addition of a base bleed unit. The object is to provide a carrier shell which will distribute a number of anti-personnel and anti-matérial bomblets over the target area. The shell body contains 72 sub-munitions – 48 x M42 grenades and 24 x M46 grenades – in the main part, which is closed at the rear by a base bleed unit. The ogive, or nose of the shell, contains a small expelling charge and the nose is closed by a time fuze.

The shell is loaded and fired in the usual way and the propellant flash ignites the base bleed unit. This burns for about 22 seconds, filling the base area with gases and thus reducing the drag and increasing the range. At the end of the time set on the fuze the expelling charge is ignited and explodes. The explosion pressure pushes down on a pressure plate at the top of the stack of grenades and this pressure is transferred through the stack to the base unit which is forced out of the shell body. The grenades are also forced out and the spin of the shell distributes them around the trajectory before they begin to fall. As they fall, they are armed and a nylon streamer ribbon is released which acts as a tail surface and ensures that they fall nose first. Each grenade consists of a shaped charge of 30g of Composition B. The M42 grenades have the inner side of their body serrated so as to break into regular fragments, while the M46 grenades have a heavier and thicker body with no control of fragmentation. Both have impact fuzes which detonate the shaped charge as soon as they strike anything. The resulting shaped charge jet is capable of penetrating 70mm of armour steel and both types deliver anti-personnel fragments in varying sizes around the point of burst.

It is also possible to use this shell as a registration round. Since it does not match any other shell in the 155mm armoury, it is not possible to use an ordinary HE shell for determining the range to the target.

When firing specialised shells of this sort, it is usual to begin by firing an ordinary HE shell so as to determine the range to the target and verify the gunnery data. If the shell fails to hit the target, more are fired, adjusting range and azimuth, until a hit is obtained, after which the specialised shell is fired. In this case, however, the M864 does not match any other shell and it has to be employed as a 'registration round'. To do this, the ogive is removed and the expelling charge is replaced by a special shaped charge. The shell is then fired on the calculated data for the target and when the fuze functions the shaped charge fires into the body and detonates all the 72 grenades at once. The resulting explosion of the shell is easily visible by observers or radar and the accuracy of the firing data can be assessed and, if necessary, corrected, before the shell is fired in its proper role.

DATA
Calibre 155mm (6.1in)
Weight 102lb (46.3kg)
Length 35.4in (899mm)
Payload 46 Grenade M42, 24 Grenade M46
Muzzle velocity ca 2,725ft/sec (830m/sec)
Maximum range ca 32,800 yards (30,000m) (depending upon model of gun)

(Left) The base bleed unit in flight.

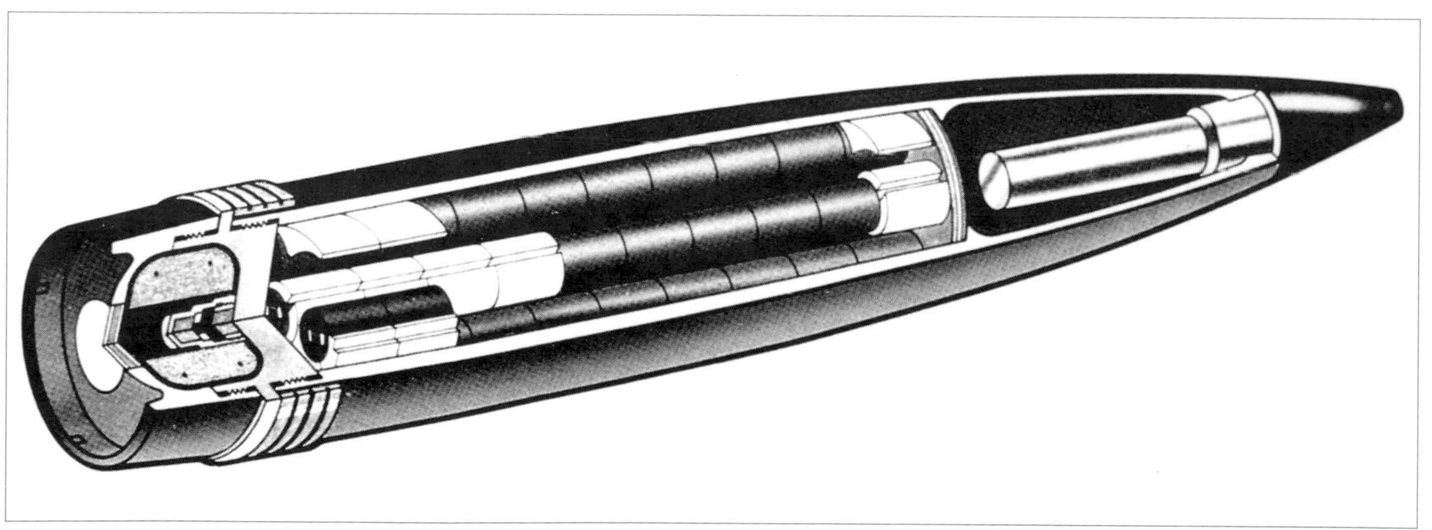

RAAM: Remote Anti-Armour Mine M1718 USA

After the original bomblet carrying shell, the M483, had entered service, the next task was to use the same system to deliver something which would inflict more damage than the small bomblets. This led to 'ADAM', the Area Denial Artillery Munition and to 'RAAM', the Remote Anti-Armor Mine. ADAM contains anti- personnel mines, while RAAM contains anti-tank mines and the significant difference between these and the bomblet shells is that the mines are larger and they do not detonate as soon as they land – they wait to be triggered.

The projectile body is of steel, with a glass fibre liner and an ogive and base cap both of aluminium. The RAAM M718 shell carries nine M73 cylindrical anti-tank mines in the body; the ogive contains an expelling charge and the nose is closed by a time fuze. The shell is fired in the normal way, the time fuze being set to burst the shell above the target area. When the fuze functions it fires the expelling charge; this blows off the shell base and ejects the nine mines which, due to the spin of the shell, spread out and fall dispersed several metres apart.

The mines are partly armed while in flight, due to the spin of the shell; arming is completed as they continue spinning after ejection. The mines land on the ground and after a short delay the fuzing system begins to function. Each mine is provided with a magnetic sensor which can detect the magnetic field generated by a tank. Once this signal is detected and reaches a strength indicating that the tank is over the mine, it will detonate. The mine is double ended, so that whichever way it lands, it will fire a curved steel plate upwards at high velocity. This plate is deformed by the force of the detonation and turns into an 'explosively formed fragment' in other words, into a slug of metal which will tear off a tank wheel or track or do severe damage to belly armour.

Should no tank stray into the area of the sensor, then after 24 hours the mine will self destruct. It will also self destruct should the internal battery power drop below a predetermined level. A number of the mines are also fitted with anti-disturbance fuzes which will discourage casual attempts to disarm or move them.

Using these RAAM shells, a six gun 155mm battery can lay a minefield 300 metres long and 250 metres wide simply by firing two rounds from each gun. By firing more rounds and by varying the elevation and direction, the field can be widened or thickened at will.

DATA
Calibre 155mm (6.1in)
Weight 102.9lb (46.7kg)
Length 33.9in (861mm)
Payload nine M73 anti-tank mines
Muzzle velocity 2,132ft/sec (650m/sec)
Maximum range 19,410 yards (17,750m)

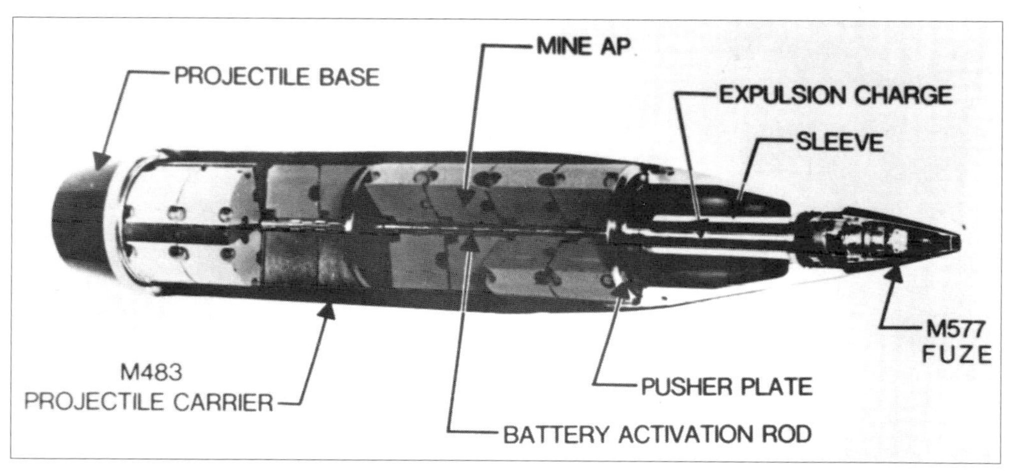
MINE AP
PROJECTILE BASE
EXPULSION CHARGE
SLEEVE
M577
FUZE
M483
PROJECTILE CARRIER
PUSHER PLATE
BATTERY ACTIVATION ROD

RAAM
PROJECTILE

155mm Cargo Shells DM642 and Rh49 — Germany

Development of this projectile began in the early 1980s, the German army requiring a bomblet shell with greater range than was then available from the American M483 design which they had adopted under NATO.

The shell body is of steel and is loaded with a total of 63 dual purpose anti-personnel and anti-armour bomblets. There is a nose ogive with an expelling charge and a time fuze and the base of the shell is recessed to reduce drag and is a push-fit held in place by shear pins. There is a broad copper driving band at the rear end of the shell body.

The bomblets are shaped charge munitions, arranged in seven stacks of nine. Each bomblet weighs 290g and is 45mm in diameter. The metal of the body is quite thick and thus, when the charge is detonated, there is a shaped charge jet for penetrating armour and a large number of high velocity anti-personnel fragments. Each bomblet has an impact fuze, but should this fail to operate then a self destruction element functions after 15 seconds and destroys the bomblet. The bomblets are provided with a nylon ribbon which stabilises them during their fall and ensures they arrive nose first and there is an aerodynamic brake on the bomblet body which greatly reduces the spin so as not to degrade the shaped charge performance.

The shell is fired in the usual manner and the time fuze is set to function when the shell is about 400 metres above the target area. The fuze then fires the expelling charge and the sudden increased internal pressure breaks the shear pins and blows off the base of the shell. The bomblets are ejected into the

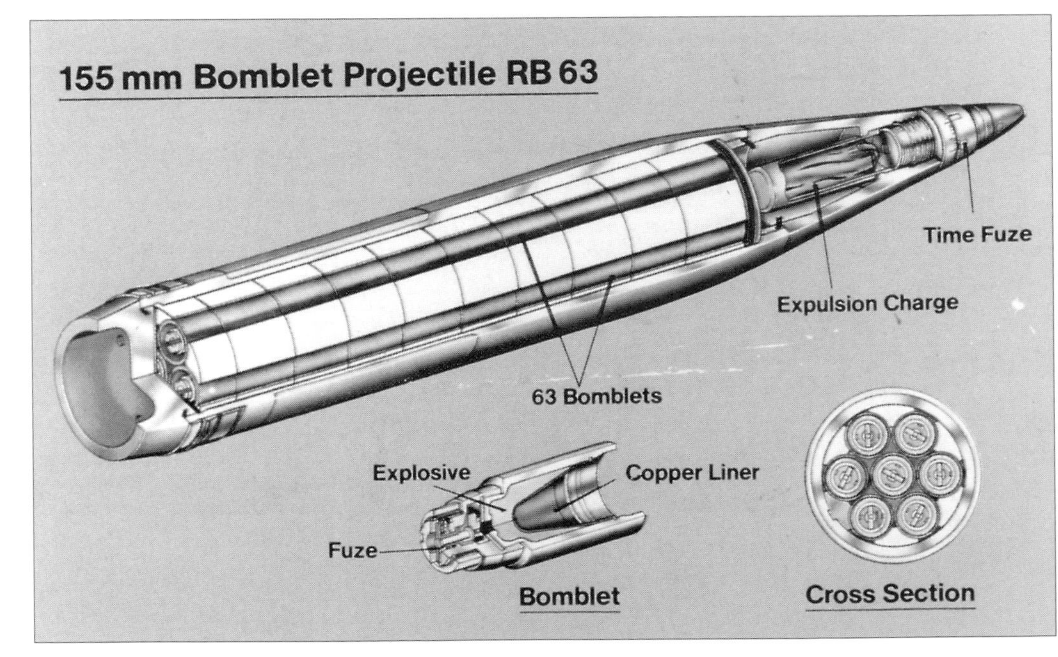

air and they are spread out by the centrifugal forces resulting from the spin of the shell. The nylon ribbon streams out and in doing so arms the fuze. The bomblets then fall to the ground and detonate on impact.

After successfully developing this projectile and seeing it into service, the Rheinmetall company then went on to develop a second version, the Rh49, which adds a base bleed unit to the rear of the shell body in place of the ordinary base. In order to compensate for the additional weight of the base bleed unit the number of bomblets is reduced to 49, but the shell functions in exactly the same manner.

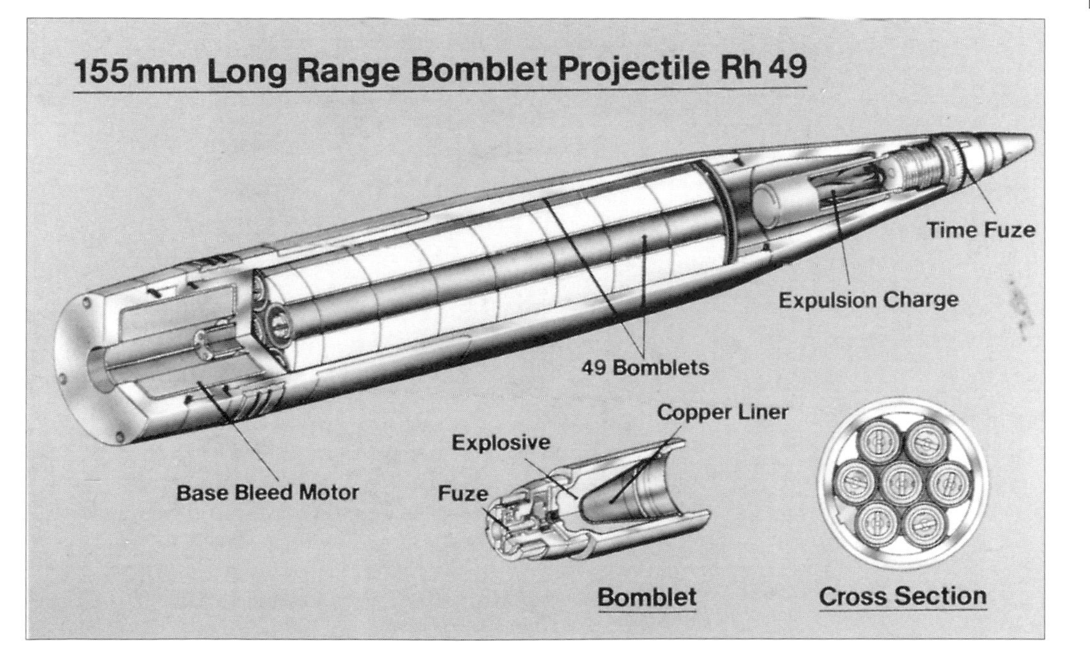

DATA
Calibre 155mm (6.1in)
Weight 1 3.6lb (47kg)
Length 35.4in (899mm)
Payload 63 bomblets weighing 27.9kg (Rh 49, 49 bomblets)
Muzzle velocity 2,631ft/sec (802m/sec)
Maximum range 24,500 yards (22,400m) (Rh49 31,165 yds/28,500m)

Glossary

The subject of ammunition has a number of words peculiar to itself; they cannot be avoided without explanations on every other page, so this glossary is intended to help explain some of the more obscure terms not fully explained elsewhere.

ARMED

A fuze is said to be armed when all the various safety devices have been freed and the mechanism will detonate the shell as soon as it strikes the target or after a set period of time. Fuzes are designed so that arming does not take place until the shell has flown some distance from the gun.

BALLISTIC CAP

A pointed light metal cap over the nose of a shell or shot to give it better flight characteristics. Some types of piercing shell for use against armour or concrete have heads designed for penetration which are not well shaped for flight and thus require a ballistic cap.

BASE EJECTION

Type of projectile which contains some payload which has to be released into the air at the target, such as smoke canisters, and which is ejected by blowing off the shell base and forcing the payload out.

BERDAN PRIMER

A primer cap for small arms cartridges, the principal feature of which is that the anvil, against which the cap is crushed by the weapon's firing pin, is part of the cartridge case.

BORE-SAFE

A fuze which is so designed that it cannot detonate the projectile whilst it is still inside the gun barrel, but must fly some distance before the fuse is armed.

BOURRELET

The machined surface at the maximum diameter of a bomb or shell. This has to be a carefully controlled dimension, to fit the weapon bore properly and, in older and simpler ammunition, the remainder of the projectile would be of lesser diameter and less well finished in the interests of economy and speed of manufacture.

BOXER PRIMER

A primer cap for small arms cartridges in which the anvil is a component part of the cap itself.

CASELESS CARTRIDGE

A small arms cartridge in which there is no brass or other metal case. The cartridge consists of a piece of explosive in which a bullet and a cap are embedded. This means that the weapon no longer has to eject an empty case, but that the weapon must provide some method of sealing the breech against the escape of gas, a job which is normally done by the cartridge case.

COMPOUND BULLET

A bullet which consists of a metal core inside a steel jacket coated with copper or gilding metal. The object is to provide a heavy bullet which will deform to enter the rifling of the weapon and will not wear away the bore by unnecessary friction or leave deposits of metal in the bore.

DISCARDING SABOT

A projectile consisting of two parts, a 'sub-projectile' which is smaller in diameter than the gun bore and a 'sabot' which is of the full bore diameter and carries the sub-projectile centred in the bore. This gives a lightweight projectile in the gun, which can reach a high velocity. On leaving the gun muzzle the sabot is flung off by various mechanisms, leaving the sub-projectile to go to the target. It allows a gun to achieve a higher velocity and hence longer range or greater penetration than can be achieved with a full calibre projectile.

GRENADE LAUNCHING CARTRIDGE

A rifle cartridge without a bullet, but with a large charge of smokeless powder. It is fired so as to provide high pressure gas to blow a grenade off the muzzle of the rifle to a distance of 100-150 yards.

OGIVE

The compound curve which forms the nose of most modern bullets and shells. The object is to blend the parallel walls into a point as smoothly as possible and with a shape which offers least resistance to the air.

SPIN

The rotational movement of a shell or bullet after being fired from a rifled gun. It is necessary in order to stabilise the projectile in flight and keep it pointed nose first towards the target: an elongated bullet fired from a smooth bore barrel will simply tumble over and over and would not follow an accurate course. The amount of spin necessary depends upon the length of the projectile and once a projectile's length exceeds about seven times its calibre, even spinning will not stabilise it.

STREAMLINED

In projectiles, 'streamlined' means a pointed nose and a tapering base, so that the airflow is led smoothly around the bullet or shell and closes behind it with the least disturbance. In American parlance, 'boat tailed' means the same thing.

STUN GRENADE

A hand grenade designed to explode and produce a loud noise and brilliant flash, but no damaging fragments, so as to disorientate and stun the target persons without injuring them. Specifically developed for use against terrorists holding hostages.

SUBSONIC AMMUNITION

Ammunition for use in silenced weapons. Silencing the blast from the muzzle is a simple mechanical problem. However, if the bullet travels above the speed of sound (1,087 ft/sec or 331 m/sec at sea level) then the sonic wave set up by its passage through the air generates a loud 'crack' which defeats the object of silencing the weapon. Subsonic ammunition sends the bullet out at less than the speed of sound, so that there is no bullet noise.

TRACER

Ammunition which leaves a visible record of its trajectory. Small arms tracer bullets have a recess in the base filled with a pyrotechnic mixture which is ignited by the powder charge as the bullet is fired. It burns during flight with an intense light, usually red, and thus draws a visible line in the air to guide the aim. Tracers on artillery shells perform the same function, but are larger. Note that phosphorus is never used in tracer ammunition; the mixture is one of barium nitrate, magnesium powder and metallic salts to give the desired colour.